KETTERING COLLEGE
MEDICAL ARTS LIBRAR

Ernö Pretsch, Jean Thomas Clerc

Spectra Interpretation of Organic Compounds

Spectroscopic Techniques:
An Interactive Course

Pretsch/Clerc
Spectra Interpretation of Organic Compounds

Bigler
NMR Spectroscopy: Processing Strategies

In Preparation:

Weber/Thiele/Hägele
NMR Spectroscopy: **Modern Spectral Analysis**

Jenny
NMR Spectroscopy: **Data Acquisition**

Fröhlich
NMR Spectroscopy: **Intelligent Data Management**

Ernö Pretsch,
Jean Thomas Clerc

Spectra Interpretation of Organic Compounds

Prof. Dr. Ernö Pretsch
Department of Organic Chemistry
ETH-Zentrum
CH-8092 Zürich
Switzerland

Prof. Dr. Jean Thomas Clerc
Gumpisbühlweg 23
CH-3067 Boll
Switzerland

A CD-ROM containing a teaching version of the program SpecTool® is included with this book. Readers can obtain further information on this software by contacting:
Chemical Concepts, P.O. Box 100202, D-69422 Weinheim, Germany.

This book and CD-ROM was carefully produced. Nevertheless, author and publisher do not warrant the information contained therein to be free of errors. Readers are advised to keep mind that statements, data, illustrations, procedural details or other items may inadvertently be inaccurate.

Editorial Directors: Dr. Christina Dyllick, Cornelia Clauß
Production Manager: Peter J. Biel

Library of Congress Card No. applied for
A catalogue record for this book is available from the British Library

Die Deutsche Bibliothek – CIP-Einheitsaufnahme
Spectra Interpretation of Organic Compounds / Ernö Pretsch ; Jean Thomas Clerc. - Weinheim ; New York ; Basel ; Cambridge ; Tokyo : VCH
(Spectroscopis techniques)
ISBN 3-527-28826-0
NE: Pretsch, Ernö; Clerc, Jean Thomas

Buch. - 1997

CD-ROM. - 1997

© VCH Verlagsgesellschaft mbH, D-69451 Weinheim (Federal Republic of Germany), 1997,
ISBN 3-527-28826-0

Printed on acid-free and low chlorine paper

All rights reserved (including those of translation into other languages). No part of this book may be reproduced in anyform – by photoprinting, microfilm, or any other means – nor transmitted or translated into a machine language without written permission from the publishers. Registered names, trademarks, etc. used in this book, even when not specifically marked as such, are not to be considered unprotected by law.
Composition: Kühn & Weyh, D-79111 Freiburg
Printing: Betzdruck GmbH, D-64291 Darmstadt
Bookbinding: Schäffer GmbH & Co. KG, D-67269 Grünstadt

Printed in the Federal Republic of Germany

Preface

Spectroscopic techniques are particularly powerful for structure elucidation when several methods are combined. An integrated view of the associated rules, reference data and tools is necessary in order to take full advantage of this synergy. The software package, SpecTool, provided on the accompanying CD-ROM makes use of present multimedia technology to present a well-balanced overview of today's spectroscopic knowledge. It provides a tightly knit navigational network based on a large number of context-sensitive links.

A set of prototype examples is used to introduce the reader into the technique of combined spectral interpretation. Wherever SpecTool can contribute to the solution the reader is prompted to exploit its possibilities. The SpecTool version supplied with this book includes all data and tools relevant to the problems presented in this book.

This volume is not an introductory textbook that provides basic knowledge of the various spectroscopic methods. It rather is intended for undergraduate students and technicians who want to gain experience in the combined application of spectroscopic methods. It will also be useful for specialists in other fields and non-chemists who want to get acquainted with the modern approach to structure elucidation. Finally, experts interested in learning about the possibilities provided by multimedia tools will also profit from this book.

Our special thanks go to Bernhard Seebass for taking care of the spectra and editing them for both the digital and analog representation as well as for careful reading of the manuscript. Dr. R. Moll, CASAF GmbH, Leipzig, inserted the digital spectra into SpecLib. We thank Prof. B. Jaun, Dr. W. Amrein, B. Brandenberger and A. Hammer for providing the spectra and Dr. Richard Knochenmuss for reading the manuscript. The careful editorial work of Dr. Ch. Dyllick and C. Clauß is gratefully acknowledged.

Working Philosophy

We believe that there is no unique approach to structure elucidation by spectroscopy that can be generalized. Depending on personal special knowledge and experience, the interpreter may start with any one spectroscopic method and try to proceed as far as possible while seeking assistance from others if he/she gets stuck or needs confirmation. Alternatively, one can proceed as we usually do: first pick the most apparent and easily accessible evidence from all spectra, put them together and then proceed further as the preliminary result suggests (creaming method). Of course, the type of structure and the extent to which structural features are reflected in the different spectroscopic data will always play a decisive role.

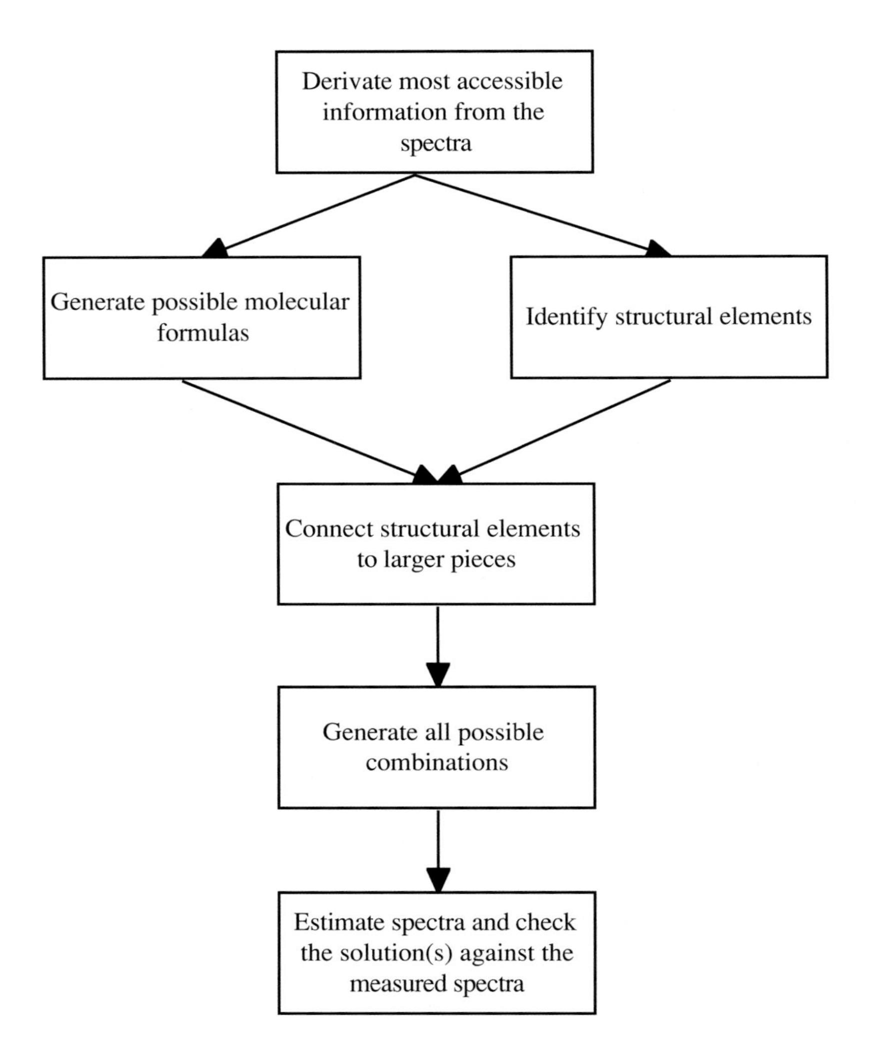

Scheme 1: General procedure of spectroscopic structure elucidation

In any case, three, sometimes overlapping, phases can be identified (see Scheme 1). In the first phase, one tries to identify substructures and structural features present (or absent) with high probability. Although the procedure varies from case to case, some general hints are proposed below. In the second phase, these building blocks are combined into full, meaningful structures. In the third phase, the spectral characteristics and features of the candidate structures are predicted and checked against the experimental evidence. If the differences are sufficiently small, the structure proposed represents one possible solution to the problem. If the fit is not close enough, one returns to steps 2 or 1. SpecTool puts heavy emphasis on step 3, and help for step 1 is provided in its Ranges section. Step 2 can be fully automated. There exist excellent programs for generating all conceivable chemical structures from a set of structural fragments and other pieces of structurally relevant information. This version of SpecTool does not include such a program.

Spectroscopic structural analysis does not always result in unequivocal answers. Even in those cases where all arguments appear to support one another and a proposed solution seems to be correct, it must be kept in mind that, in general, the analysis is terminated long before an exhaustive treatment can be claimed. Consequently there is always a finite possibility (even though at times it may be very remote indeed) that another structure, which one has not thought of, may satisfy all experimental data as well. The probability of such an event is very much smaller in combined application of several spectroscopic methods than if one relies on only one or two of them, but it can never be entirely ruled out. In any case one should minimize chances of pitfalls of this kind by carefully screening all available data for contradictory or ambiguous evidence and by trying to rationalize all those features which appear unusual or which are not apparent from the proposed structure.

The pieces of information typically searched for in the various types of spectra are as follows:

Mass spectrometry

- tentative assignment molecular mass
- occurrence elements form nominal mass and isotope distribution
- type of compound from ion series and intensity distribution
- structural elements form neutral losses (mass differences of the first fragments relative to the molecular ion)

Infrared spectroscopy

- identification of functional groups from typical absorbances (OH, NH, special types of CH, groups with triple bonds and cumulated double bonds, C=O, C=C, aromatics). Except for the most simple cases, only very intensive signals should be interpreted in the fingerprint region (1500 - 1000 cm^{-1}) at this stage.

Proton NMR spectroscopy

- integrated intensities, possible total numbers of protons
- chemical environment of protons from the chemical shifts
- neighboring groups from first order splittings
- symmetrical second order spin systems

Carbon-13 NMR spectroscopy

- number of signals: possible minimal number of carbon atoms
- number of protons attached to carbon atoms
- euivalent carbons from comparison with the integrals in the proton NMR spectrum and the presence of X–H ($X \neq C$) groups from the IR spectrum
- chemical environment of carbons from the chemical shifts

UV/vis spectroscopy

- identification of delocalized π-electron systems. UV/vis spectroscopy has today only a marginal relevance in structure elucidation work. For this reason, it is only used in a few cases and only tabulated data are given

The first 15 chapters of this volume present selected spectroscopic problems. The initial examples are quite straightforward, whereas the later ones become more and more involved. It is advisable first to try to extract as much information as possible from the spectra and to start reading the text only if one has solved the problem or gets stuck at any stage. One of many different possibilities of deriving the structure is presented in every example. It is not better in any way than any other procedure leading to the same result. Ideally, you should look at all spectra in a parallel way, however only a consecutive discussion is possible in a book.

The book can be read without using SpecTool. However, to get acquainted with the many different aspects of the enclosed demo version, the use of a computer is advisable. A more detailed instruction is given in the first chapter and only brief hints in the later ones. In the next section of this Preface instructions are given how to install and start SpecTool. Ideally this should be enough to get you started, but if you want to learn more about this program you will find a description in Chapter 16.

This volume assumes some basic knowledge of the principles of the spectroscopic techniques applied, i.e., mass spectrometry, NMR, IR and UV/vis spectroscopy. A general introduction as given in any textbook covering instrumental analysis should be sufficient. A selection of some references that provide a brief introduction to all of the methods is given at the end of the Preface. Chapters 17 - 19 provide further comments on the individual methods with emphasis on topics which, according to the author's experience often lead to difficulties.

How to install SpecTool

Run the install.exe program from the CD-ROM and follow the guidelines given on the screen. In the SpecTeach version, SpecTool runs on Windows 3.1, Windows 95, and Windows NT, and requires about 28 MB disk space. The install procedure will copy the font file and create a program group. SpecTool fonts become effective only after Windows is restarted. If the program manager fails to create a program group, a link to the program has to be built manually. The program call is TMBOOK.EXE with SPT.TBK as command line parameter. Both these programs can be found in the SpecTool directory. SpecTeach should be run with the SpecTool directory as the current directory.

To erase SpecTeach from the system, delete the SpecTool directory with all its subdirectories, delete the program group, and use the Control Panel to uninstall the fonts.

Selected References

R. M. Silverstein, G. C. Bassler, T. C. Morrill: Spectrometric Identification of Organic Compounds, 5th Edition,Wiley, New York, 1991.

L.D. Field, S. Sternhell, J.R. Kalman: Organic Structures from Spectra, 2nd Edition, Wiley, New York, 1995.

D. H. Williams, I. Fleming: Spectroscopic Methods in Organic Chemistry, 5th Edition, McGraw-Hill, London, 1985.

Contents

1 Problem 1

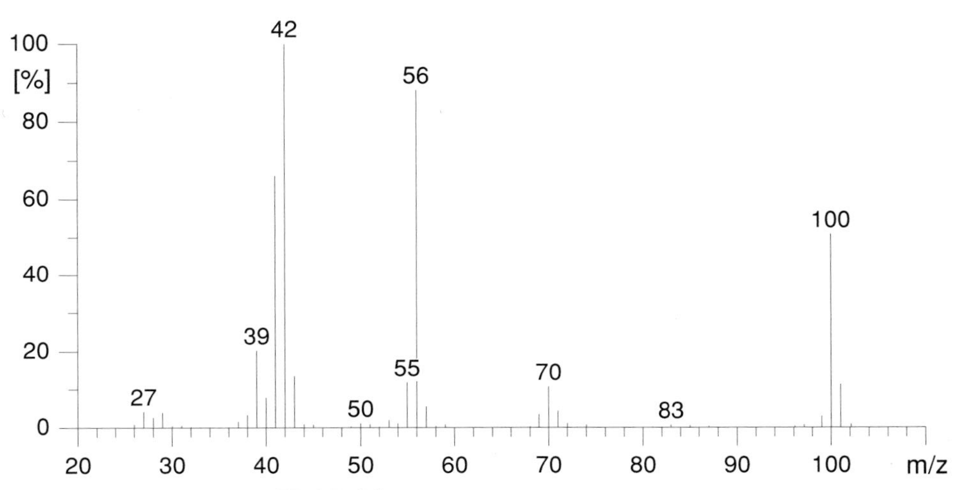

Fig. 1.1: Mass spectrum: EI, 70 eV

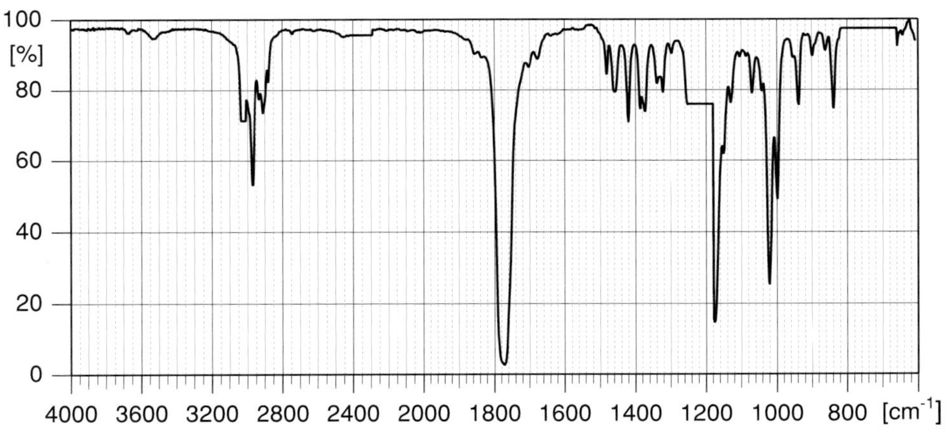

Fig. 1.2: IR spectrum: recorded in $CHCl_3$, cell thickness 0.2 mm

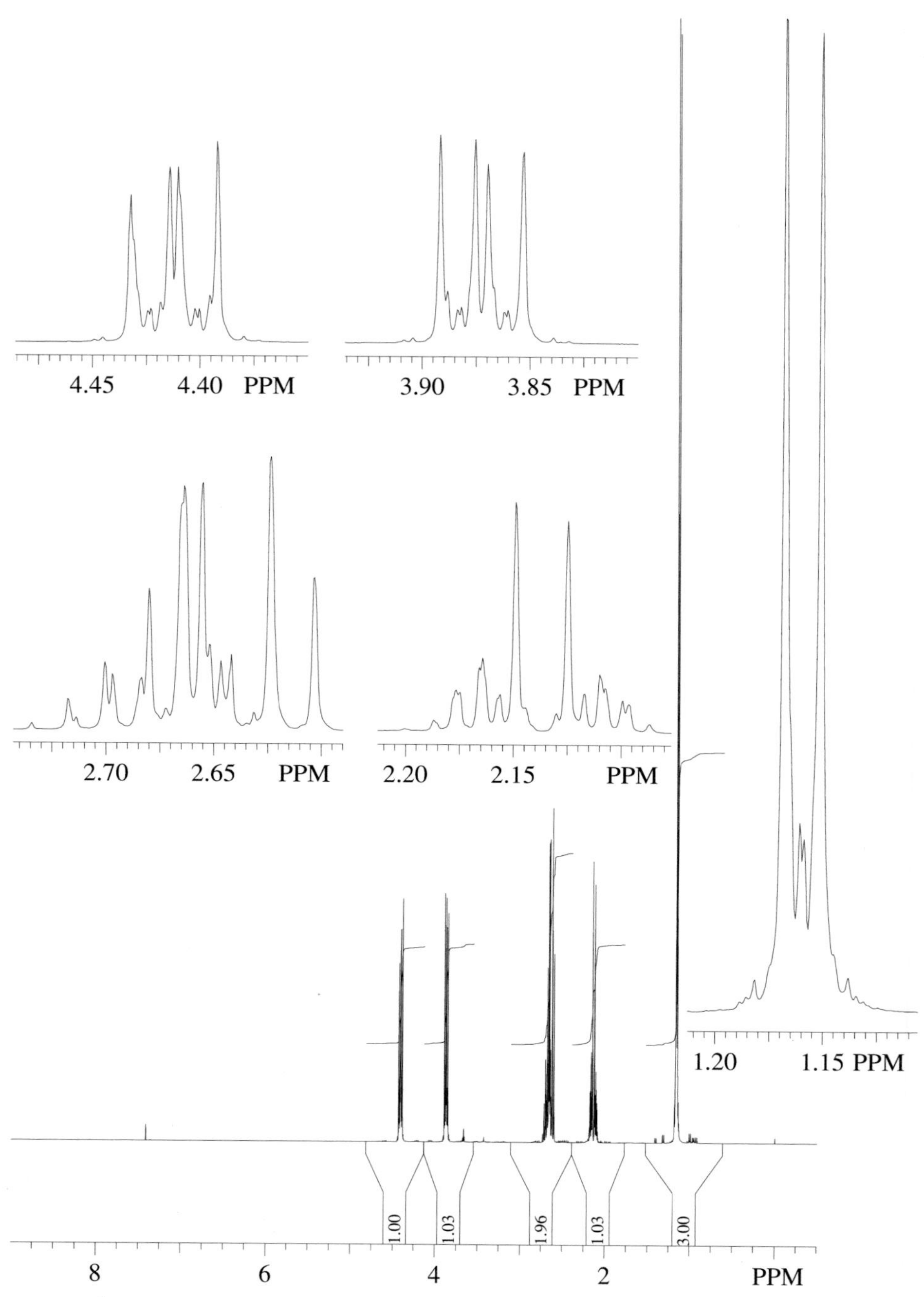

Fig. 1.3: ^{1}H-NMR spectrum: 400 MHz, solvent: $CDCl_3$

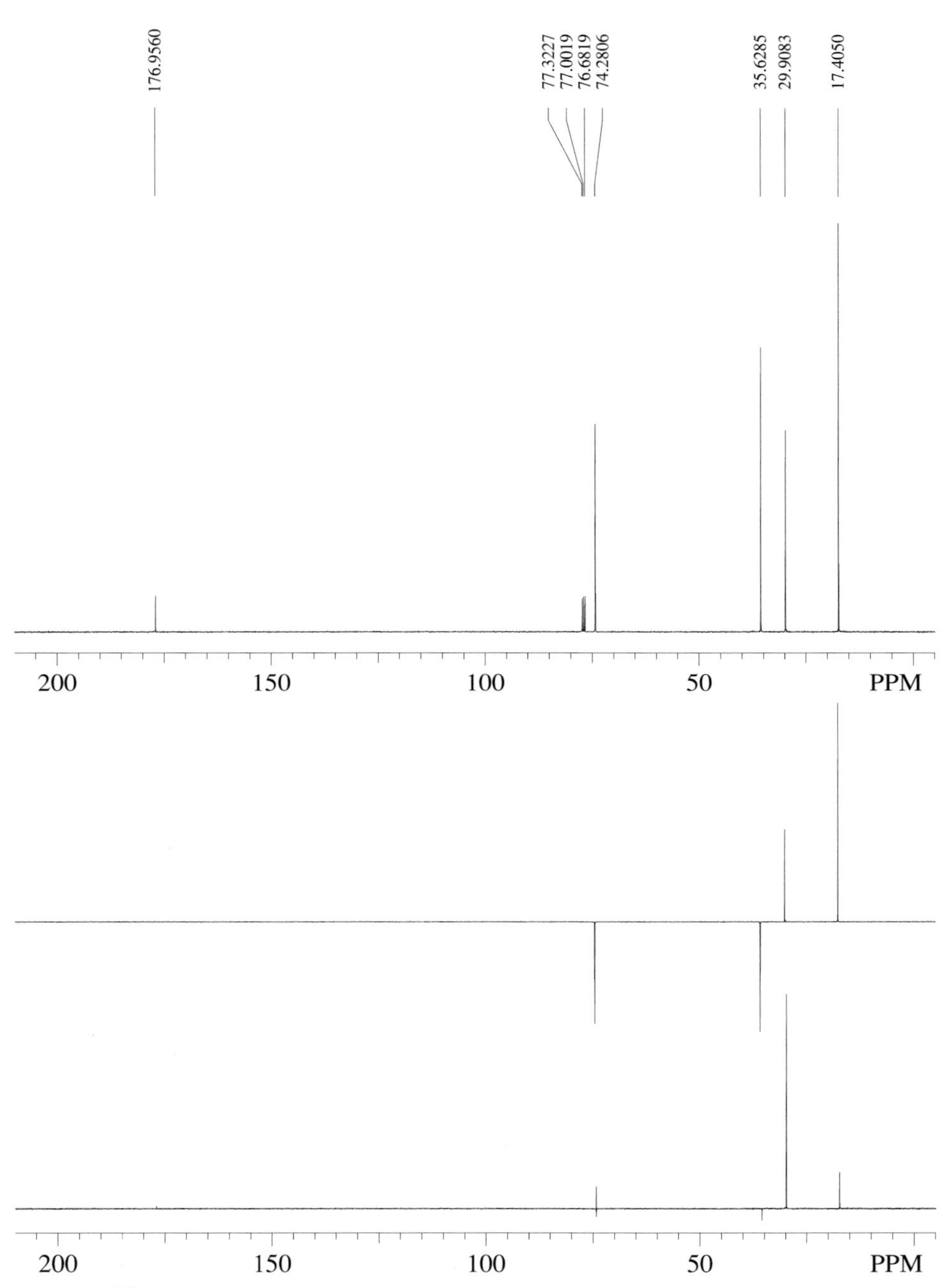

Fig. 1.4: ^{13}C-NMR spectrum: 100 MHz, solvent $CDCl_3$, Top: proton decoupled, middle: DEPT135, bottom DEPT90 ($\tau = 3.6$ ms)

1.1 Elemental Composition and Structural Features

In the ^{13}C-NMR spectrum the signals at 17.4, 29.9 and 35.6 ppm can be assigned on the basis of the DEPT spectra to a methyl, methine, and methylene group, respectively (cf. Comments). The signal at 74.3 ppm indicates a methylene group, which is most probably attached to an oxygen atom as indicated by the large chemical shift. Chemical shift values of 60 ppm or higher for aliphatic methylene groups generally suggest an oxygen atom as substituent if no halogen atoms or nitro groups are present in the molecule.

Check it in SpecTool:
If you are not already on the **TOP** page, go there by selecting **TOP** page from the **Page Menu**. On the **TOP** page, select **CNMR Data**, and then select **Aliphatic Compounds**. This can be done by one of the three following operations:

- Click at the **Alkane** symbol on the displayed page (upper left corner),
- Click at the **Alkane** symbol on the **Grp** (Group) palette,
- Select **Alkanes** under the **Groups Menu**.

Choose the monosubstituted alkanes by clicking at the appropriate button. A table is now shown with the chemical shifts of substituted methyl, ethyl, propyl, isopropyl and t-butyl groups. This table provides a fast overview of the effects of the most common substituents. You can rearrange the table according to the chemical shift of one group by clicking with the mouse at the corresponding heading. The distribution diagram at the bottom of the page shows the spread and density of the data. Clicking on the diagram marks the nearest entry and scrolls it into view.

The signal at 177.0 ppm indicates a O=C–X group, X being a hetero atom other than sulfur, otherwise the chemical shift would be considerably higher.

Check it in SpecTool:
On the **Grp** palette, showing the various substructures, you select the last item **Sp**, standing for Special. (This will not work from the **Top page** as no method is defined.) From the target node, **Special Data**, reference data can be accessed which cannot be assigned to one single compound class. By selecting the rightmost button in the top row (**C=O**) a card with the chemical shifts of various carbonyl compounds is presented.

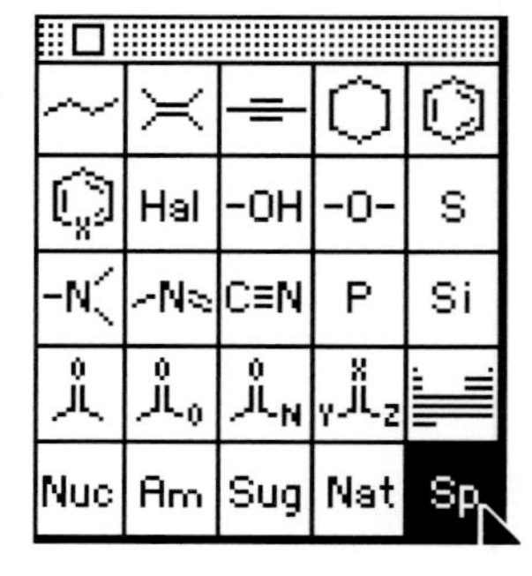

The elemental composition of the structural elements found so far amounts to C_5H_8O (mass 84).

The integration of the signals in the proton NMR spectrum gives the following intensity ratios: 1 : 1 : 2 : 1 : 3 and corresponds to a total of n × 8 protons (n= 1, 2...). The splitting of the methyl signal indicates a CH_3–CH group as structural fragment.

Check it in SpecTool:
Access the ^{1}H-NMR spectrum in **SpecLib**, by clicking at the corresponding button (column: HNMR row: SpecLib) on the **Top** page and at **Problem 1** on the then appearing page. Zoom in into the various parts of the spectrum by selecting the starting point with the mouse, and dragging to define the required segment. There is an Unzoom entry in the Edit menu. Be sure to close all SpecLib Windows before clicking the button "Back to SpecTool".

As all fragments correspond to chemically reasonable mass differences relative to the last significant signal at m/z 100 in the mass spectrum, we assume a molecular mass of 100. The difference between this molecular mass and the mass of the structural fragments found so far amounts to 16 mass units and must be assigned to an oxygen atom, since no additional protons are available. The molecular formula thereby becomes $C_5H_8O_2$ and shows two double bond equivalents. One of them is taken care of by a carbonyl group, the other one must be due to a ring since no other double bond is indicated.

Check it in SpecTool:
The **MolForm** program is made accessible by selecting **General Tools** from the **Top** page (or by changing the method to **MS** and the topic to **Tool** by using menus or palettes). It calculates all possible molecular formulas for a given mass range which are compatible with the entered element ranges (and/or weight percentages). Try it by clicking the appropriate button. This starts **MolForm** which is an external program and runs independently of the **SpecTool** kernel. Enter 100 into the **Mass** field (for low resolution spectra the nominal mass with the default tolerance of 0.5 mass units can be used at masses not higher than about 500) and run the program. It will suggest 17 possible molecular formulas. By clicking at any of them further information is displayed, including the exact (monoisotopic) mass. Narrowing the bounds efficiently constrains the possibilities. Here we can enter the number of hydrogen atoms as being 8. Set Min = Max = 8 and try again. To end the program, click the exit button or select **Close** from the system menu and then click the **Back to SpecTool** button.

The presence of a carbonyl group is confirmed in the infrared spectrum by the C=O stretching vibration band at 1770 cm^{-1}. The strong absorption band at 1170 cm^{-1} is very likely to be due to a C–O–C stretching vibration. The range of ca. 1200 - 1260 cm^{-1} is biased by solvent absorption.

Check it in SpecTool:
The infrared spectra are not reliable in ranges where the solvent absorbs strongly. In the **IR Spectra** section you can visualize these ranges. Access the **Top** page e.g., by clicking at the page title. This brings up a menu of the logical precursors of the current page (here just **Top**). Clicking at an entry brings you to the respective page.
The **Spectra** section includes selected reference spectra whereas **SpecLib** contains the spectra of the compounds discussed in this volume which can be zoomed and manipulated in various ways.
Select **IR Spectra** and navigate to the spectrum of an arbitrary ester.

Use **Show opaque regions of solvents**, then **Solvents** and select **$CHCl_3$ 0.2 mm cell thickness**. Regions biased by solvent absorption are indicated by black bars.

1.2 Structural Assembly

The structural fragments found above may be put together in only two different ways:

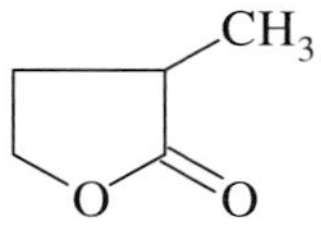

CH_3 O O CH_3 O O

I 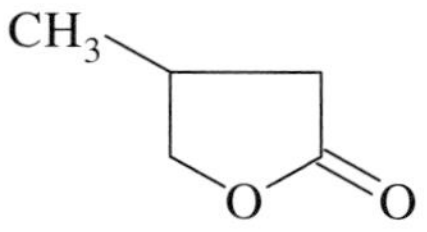 **II**

The constitution **I** may be eliminated by the following reasoning: The large chemical shift difference for the diastereotopic CH_2–O protons (3.87 and 4.41 ppm) can only be rationalized if the methyl substituent is attached to the neighboring ring carbon atom (the assymmetry center, see also below).

1.3 Comments

1.3.1 Mass Spectrum

The mass spectrum illustrates the fact that occasionally even mass fragments can dominate in the absence of nitrogen, if the compound consists to a large extent of especially good leaving groups like CO and CO_2. The choice between **I** and **II** could safely be based on a mass spectrometric evidence if it was shown by accurate mass measurement that m/z 42 is mainly due to ketene ions ($CH_2C{=}O$, in **II**) rather than C_3H_6 (in **I**).

Check it in SpecTool:
The module **Isotope Pattern** can be used to check the necessary resolution of the mass spectrometer. Navigate e.g., via the **Top** card to the **MS Tools** and start it. Set **Formula** to C2H2O and **Run** to get 42.010 for the lowest monoisotopic mass. Repeat the procedure for C3H6, note the mass difference and estimate the necessary resolution. Use close from the system menu to exit the program and then click at the "Back to SpecTool" button.

1.3.2 Infrared Spectrum

The most conspicuous feature of the infrared spectrum is the high frequency of the carbonyl stretching vibration. Carbonyl stretching vibrations are shifted to higher frequencies primarily by decreasing the angle between the substituents, and by substitution with electronegative groups. The world record in carbonyl stretching frequency is held by carbonyl fluoride COF_2, which absorbs at 1928 cm^{-1}. In the compound at hand, it is the steric influence of the five-membered ring on the C-O-CO-C moiety that is responsible for the high frequency.

> **Check it in SpecTool:**
> Visit the IR data page **Lactones** by selecting in sequence the **Top**, **IR-DATA**, **COO** and **Lactones** buttons. The target card schematically shows the most important absorption bands of this compound class. Click at a band to display reference data. Use **Back** from the **Page Menu** to return.

In infrared spectra, the light absorption is commonly recorded as transmittance, expressed in %. Transmittance is the ratio between the intensity of the transmitted light in the sample beam to the intensity of the reference beam. According to Beer's law, transmittance ideally varies exponentially with the concentration c, cell thickness l, and extinction coefficient ε:

$$T = \frac{I}{I_0} = 10^{-\varepsilon c l} \qquad (1.1)$$

Displaying in absorbance $A = -\log T = \varepsilon c l$ becomes increasingly popular because it is directly proportional to concentration and path length. **SpecLib** allows you to switch between the two modes.

> **Check it in SpecTool:**
> Try the various display modes with the spectrum of this Problem. Use Edit/SpecLib Preferences/IR to change the settings. The new ones become effective when you click into the window.

1.3.3 Proton NMR Spectrum

The protons of the CH_2–O group are diastereotopic and may thus be at best accidentally isochronous. Here, the difference of the corresponding chemical shifts is ca. 0.6 ppm. This large difference is due to the vicinal methyl group. The influence of the methyl groups on the chemical shift of protons in γ-position is strongly dependent on the conformation. A synplanar conformation leads to a shielding effect of ca. 0.8 ppm and an antiplanar conformation causes a deshielding of the same proton by ca. 0.3 ppm:

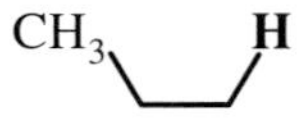

CH_3 **H** $\delta = -0.8$ relative to H **H**

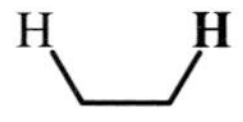

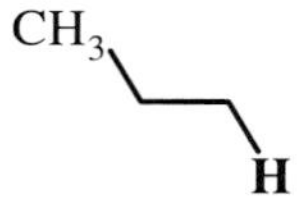

CH_3 **H** $\delta = +0.3$ relative to H **H**

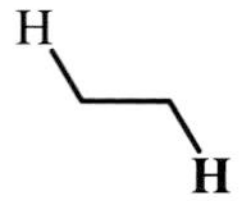

Correspondingly the CH_2CO protons can be assigned to the signals at approximately $\delta = 2.14$ and 2.64 ppm while the methine proton has a chemical shift close to $\delta = 2.67$ ppm. Chemical shift arguments result thereby in an easy assignment of each one of the geminal methylene protons. No such assignment would be possible by considering vicinal coupling constants, very much in contrast to the situation encountered in case of a six-membered ring (see Problems 6 and 15). In five-membered rings, because of their greater flexibility no simple correlation exists for the relative magnitudes of the *cis*- and *trans*- coupling constants.

1.3.4 Carbon-13 NMR Spectrum

In ideal cases, the interpretation of DEPT (acronym for "Distortionsless Enhancement by Polarization Transfer") spectra is very simple: CH_3 and CH carbon atoms as positive, and CH_2 as negative signals in the DEPT135 spectrum. Only CH signals are detected by the DEPT90 experiment. No signals appear in DEPT spectra for carbons, not directly attached to a hydrogen atom. However, multipulse techniques for identifying the number of protons attached directly to a carbon atom assume one single fixed value for all $^{13}C-^{1}H$ one-bond coupling constants. If a real value deviates considerably, additional, generally small signals may appear (cf. DEPT90 spectrum of this example, and Comments of Problem 7). If the deviation is huge the signal might be outright wrong, e.g., in case of acetylenes.

2 Problem 2

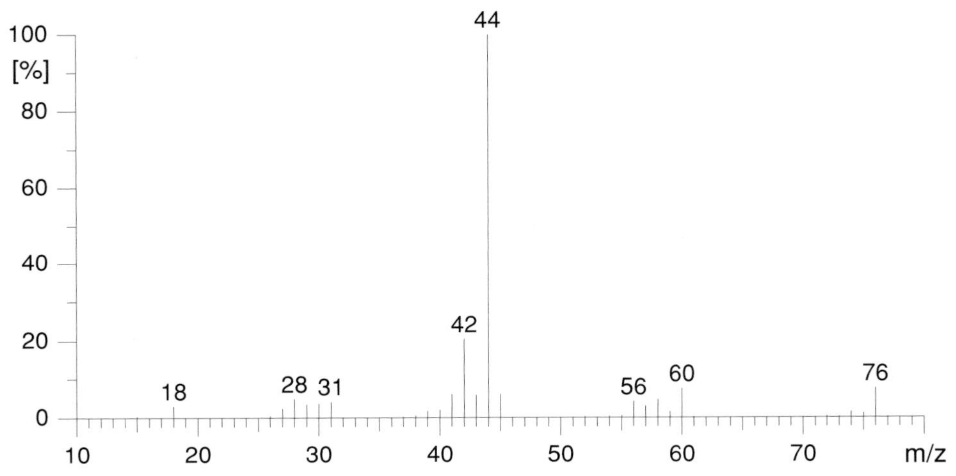

Fig. 2.1: Mass Spectrum: EI, 70 eV

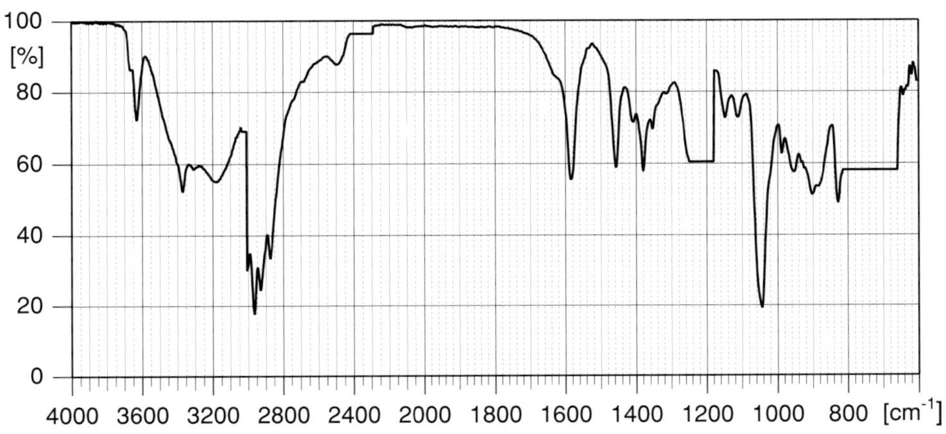

Fig. 2.2: IR Spectrum: recorded in $CHCl_3$, cell thickness 0.2 mm

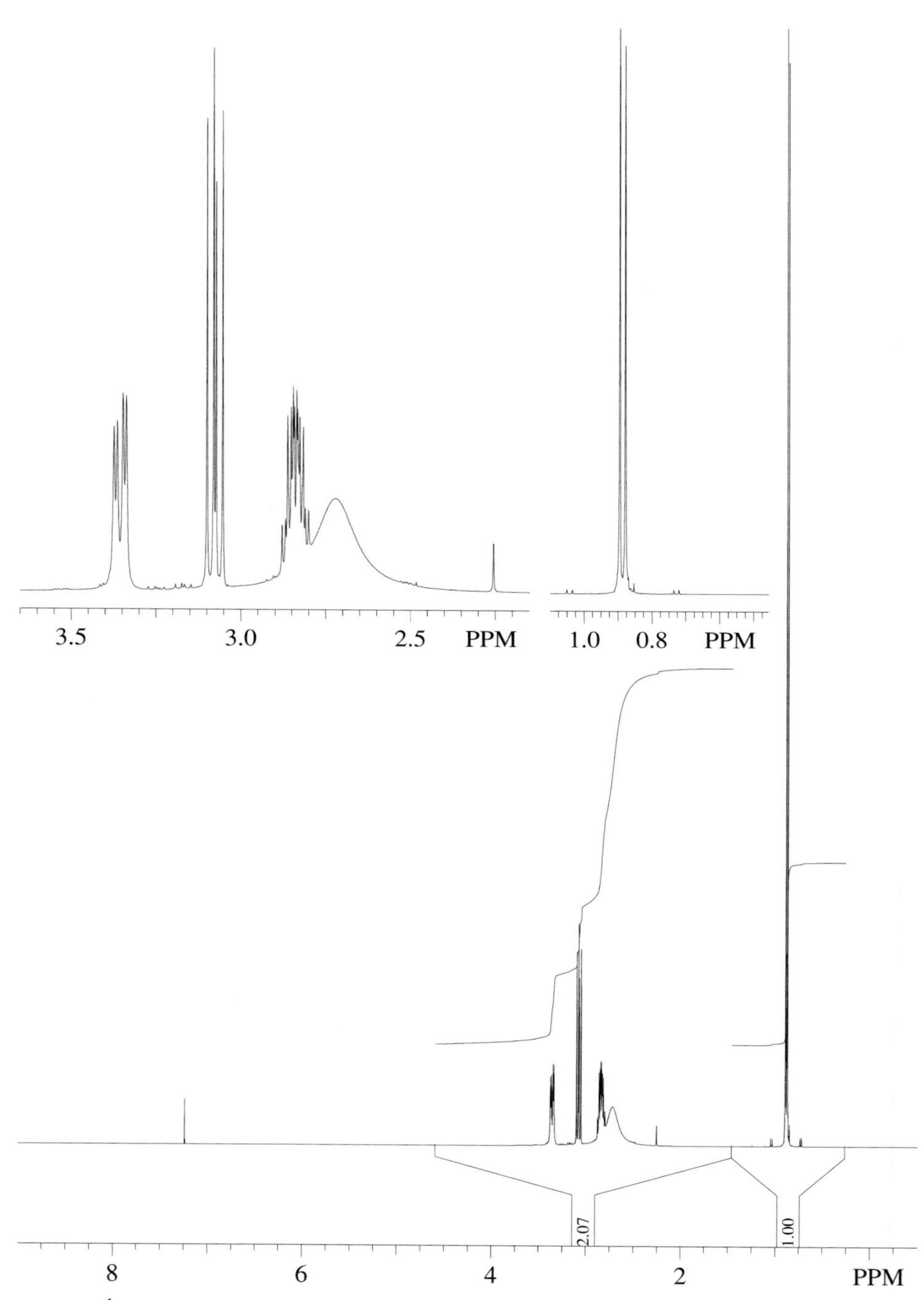

Fig. 2.3: ^{1}H-NMR spectrum: 400 MHz, solvent: $CDCl_3$

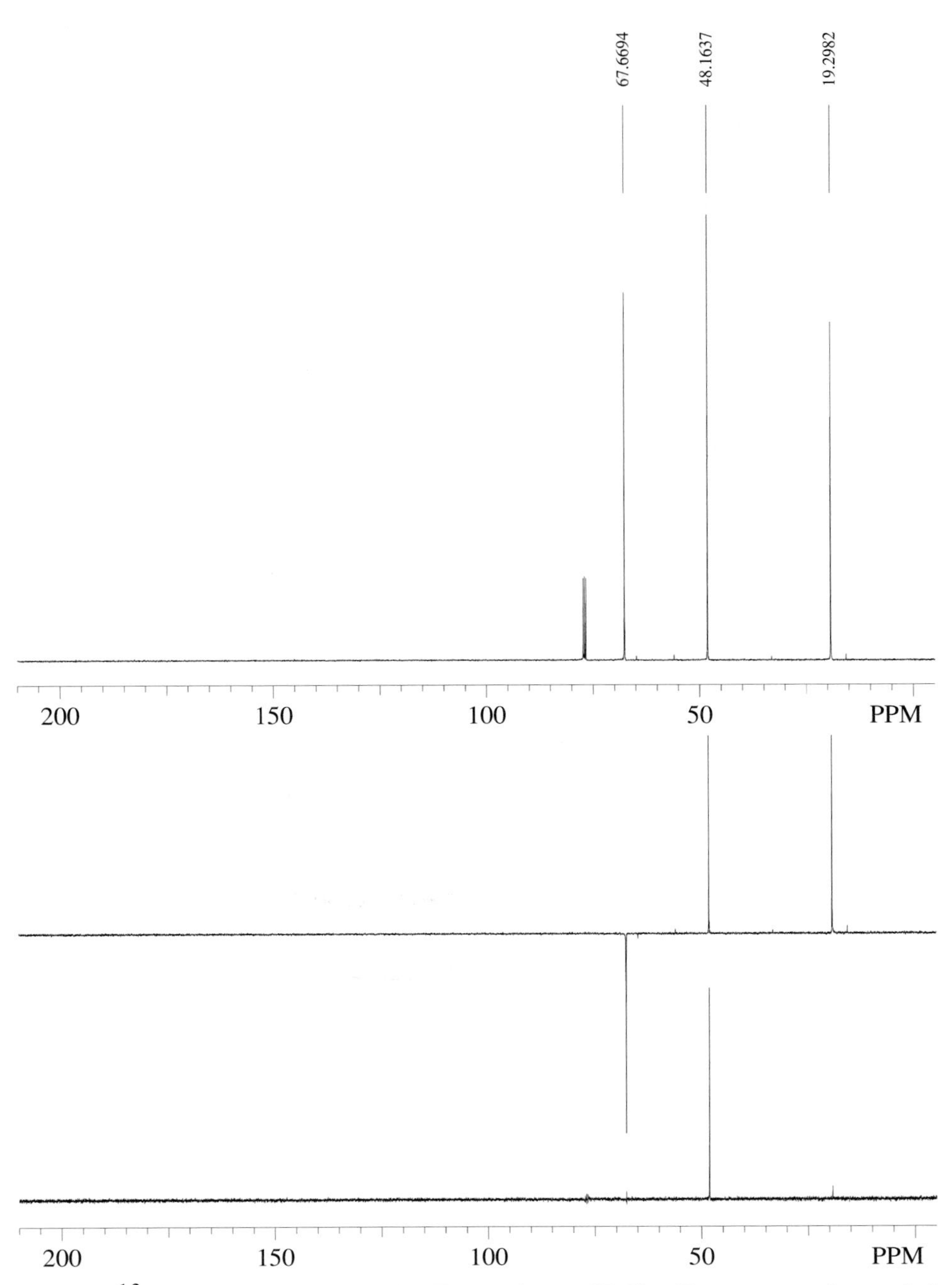

Fig. 2.4: ^{13}C-NMR spectrum: 100 MHz, solvent $CDCl_3$, Top: proton decoupled, middle: DEPT135, bottom DEPT90 ($\tau = 3.6$ ms)

2.1 Elemental Composition and Structural Features

The relatively sharp band in the infrared spectrum at 3630 cm^{-1}, together with the broad bands extending from above 3500 cm^{-1} to 2400 cm^{-1} indicate a partially hydrogen bonded hydroxyl group. The former band is assigned to the free hydroxyl group, whereas the latter corresponds to the associated form. An amino or imino group, if present, would be almost completely masked by the hydroxyl absorption. However, the small maximum at 3370 cm^{-1} might be taken as indication of an NH (or NH_2) group, being due to the stretching vibration of the (non associated) NH. No obvious assignment is possible for the band at 1590 cm^{-1}. Due to its shape, skeletal vibrations from an aromatic ring can be excluded while its intensity is too low for a carbonyl band. If NH is indeed present, it might be its deformation vibration.

Consult SpecTool:
Information on the various relevant vibrations of the NH_2 and NH groups can be accessed from the **Top** card by selecting **IR Data**, **N(sp^3)** and **Primary** respectively **Secondary Amines** in sequence. Click at a selected band or in a table on these pages for further information.

In the mass spectrum, the peak corresponding to the highest mass occurs at m/z 76. This cannot be the molecular ion, as the mass difference to the next important peak at m/z 60 is 16, which would correspond to methane, oxygen, or NH_2. Formation of these fragments from the molecular ion is, however, not very likely.

Check it in SpecTool:
To check the significance of signals in the mass spectrum select **MS Ranges** on the **Top** card. Set the cursor on the mass scale to 16 and click. The data displayed show that the loss of 16 is limited to special cases.

We further note that the mass spectrum is dominated by ions at m/z 44 and 42.

Consult SpecTool:
Inspecting these three masses in **MS Ranges** do not allow for a clear conclusion at this stage.

Integration in the proton NMR spectrum gives 2 : 1 and can be further refined to 1 : 1 : 4 : 3, the doublet at δ = 0.88 obviously corresponding to a methyl group connected to a methine group which has its signal around δ = 2.84. This proton is further coupled with other protons which resonate at 3.1 and 3.4 ppm. The broad signal at δ = 2.7 has to be left unassigned for the moment. Taking into account that the signal at 2.84 ppm corresponds to a methine, it contains three protons, including the one(s) from the hydroxy group(s) inferred from the infrared spectrum.

Consult SpecTool:
Access the ^{1}H-NMR spectrum of this Problem in **SpecLib**, by clicking at the corresponding button (**HNMR – SpecLib**) on the **Top** page and at **Problem 2** on the then appearing page. Zoom in into the various parts of the spectrum and try to

assign coupling partners. The distance (in Hz) between two signals can be displayed at the top of the spectrum to the left of the chemical shift. First, the cursor is set at the starting point and the SHIFT key is pressed, upon which aan arrow appears marking this position. As long as the SHIFT key is held down when moving the cursor, the distance from the starting point will be shown.

The carbon-13 NMR spectra confirm the deductions from the proton NMR spectrum. We have the methyl group at $\delta = 19.3$ and the neighboring methine group at $\delta = 48.2$. Its other coupling partner in the proton NMR spectrum must be the methylene group at $\delta = 67.7$. Due to the chemical shift value both these groups have to be vicinal to a hetero atom. We have thus identified the following partial structure:

$$\left.\begin{array}{l} CH_3 \\ | \\ CH-Y \\ | \\ CH_2-X \end{array}\right\} 3H$$

2.2 Structural Assembly

One of the two hetero atoms has to be an oxygen, as the infrared spectrum requires a hydroxyl group. This atom must be connected to the methylene group according to the chemical shift value of $\delta = 67.7$ in the carbon-13 NMR spectrum (see Comments). As the other hetero atom has to accommodate the two remaining protons, it is most likely a nitrogen atom in a primary amine group. Thus the constitution of the unknown compound is:

$$\begin{array}{l} CH_3 \\ | \\ CH-NH_2 \\ | \\ CH_2-OH \end{array}$$

The molecular mass is 75 mass units. The last peak in the mass spectrum at m/z 76 is thus due to the protonated molecular ion. Protonated molecular ions are commonly observed for primary aliphatic amines. The unassigned band at 1590 cm^{-1} in the infrared spectrum may now also be explained as being due to the deformation vibration of the primary amino group.

2.3 Comments

2.3.1 Mass Spectrum

The fact that even mass fragments dominate the mass spectrum and the presence of a significant peak at m/z 30 could be taken as primary diagnostic evidence for the presence of nitrogen. A very intense fragment of even mass within the nitrogen series m/z 30, 44, 58, ... is always suggestive of a saturated aliphatic amine residue.

Consult SpecTool:
The **MS Data** card **Primary Alkylamines** under **Nsp^3 Data** gives more detailed information including a prototype spectrum and fragmentation schemes. Quick scrolling in the text sections is provided by buttons at the bottom bar of the page.

Analogously, the presence of a significant m/z 31 is a reliable indicator for singly bonded oxygen.

If a spectrum (as the present one) ends with a cluster of weak signals, the assignment of a specific molecular mass becomes more ambiguous, because evaluation of isotope peak intensities is difficult or impossible. The consideration of the mass differences between all observable fragments becomes more critical and one performs a trial and error analysis in order to find the most convincing solution. Since, apart from hydrogen, 15 (CH_3) is the smallest mass difference, which is chemically reasonable and, in addition, loss of CH_3 radical is by far the most common primary fragmentation, the first attempt in such an analysis will be assumption of a molecular mass 15 units higher than the largest significant fragment mass, in this case 60 + 15 = 75. Under this assumption, the rationalization of the signal cluster around m/z 75 becomes very reasonable, because m/z 76 is to be interpreted as protonated molecular ion and m/z 74 as product of a deprotonation reaction, while m/z 58, 57, and 56 become water elimination products.

Consult SpecTool:
The **MS Data** card **Aliphatic Alcohols** under **Alcohols** gives more detailed information including a prototype spectrum and fragmentation schemes.

Protonation of the molecular ion is a highly probable process anyway if aliphatic nitrogen and hydroxyl functions are indicated by other evidence. The tendency to protonate is especially pronounced in aliphatic amines, nitriles and esters, though other polar groups are subject to such reactions and also often exhibit higher intensities of first isotope peaks of their molecular ions than required by elemental composition and natural isotope distribution. Assuming m/z 76 to represent the molecular ion results in less satisfactory explanations for the spectral features. A mass difference of 16 units to the first significant fragment is chemically not entirely unreasonable, but restricted to a few specific types of structures. In addition loss of one and two hydrogen atoms with equal probability would be a quite unusual feature. The presence of m/z 18 (ionised water) should not be used for structural argument, regardless of its intensity, because water is always present adsorbed on the sample or instrument surfaces and cannot be differentiated from water formed by degradation of the investigated compound.

2.3.2 Infrared Spectrum

For molecules, as simple as the one which we have here, it is often possible to assign most major infrared absorption bands by using correlation tables and reference spectra.

At 3630 cm^{-1} we have the OH stretching vibration of the free hydroxyl group. The free NH_2 group exhibits two stretching vibrations, one for the asymmetric mode at 3370 cm^{-1}, the other for the symmetric mode is expected at 3300 cm^{-1}. There are indeed very small peaks discernible at these frequencies. The vibrations of the associated OH as well as of the NH_2 group give rise to the broad band from 3600 cm^{-1} to 2400 cm^{-1}. It is uncommon for simple alcohols and amines to have the stretching vibrations of the associated species extending much below 3000 cm^{-1} in dilute solutions. In the present case, this is most probably due to the formation of intramolecular as opposed to intermolecular hydrogen bonds. Another possibility is that the amine group is to some extent protonated. The anion may be chloride formed by decomposition of chloroform or, carbonate formed by reaction with carbon dioxide and moisture from the ambient air. The group of bands between 3000 cm^{-1} and 2800 cm^{-1} is, of course, due to the various CH stretching modes. At 1590 cm^{-1} we have the NH_2 deformation vibration, as already stated in the foregoing. At 1460 cm^{-1} we find CH_2 deformation and CH_3 asymmetric deformation vibrations. The CH_3 symmetric deformation absorbs at 1380 cm^{-1}, and the low intensity band at 1350 cm^{-1} is ascribed to the deformation of the CH group bearing the nitrogen atom. From 1270 cm^{-1} to 1200 cm^{-1} the spectrum is masked by solvent absorption and is thus not interpretable. Skeletal vibrations including CN and CO stretching modes are expected in this region. The strong absorption at 1040 cm^{-1} is most probably due to CO stretching. Below 1000 cm^{-1} we have various bands of lesser intensity ascribed to OH deformation.

In those spectral regions where the solvent exhibits a strong absorption, little light is received by the detection system of the spectrometer. The apparent transmittance is therefore unpredictable and depends primarily on parameters not under control of the operator. The recorder pen may drift to either side or may remain stable. Moreover, it may also follow an erratic trace which may by chance have the appearance of a real absorption band. In most cases, however, regions of strong solvent absorptions are readily identified by unnatural looking band shapes. Single beam Fourier transform instruments usually insert a horizontal line in these regions. This is well exemplified in the present example for the bands at 1250 cm^{-1} and between 810 cm^{-1} and 650 cm^{-1}.

2.3.3 Proton NMR Spectrum

The two amine and one hydroxy protons give rise to a broad line at $\delta = 2.7$. Usually the exchange between amine and alcohol protons is fast on the NMR time scale and the protons show equivalence by a kinetic mechanism. Strong hydrogen bonding presumably contributes to the somewhat high value of the observed weighted average of the chemical shift values.

Since the molecule is chiral, the two methylene protons are diastereotopic and part of an almost first order A_3XYZ spin system. The vicinal coupling constants of the two diastereotopic protons to the methine proton are quite different indicating a preferred conformation, probably stabilised by intramolecular hydrogen bonds.

The difference of the heights of the two 4-line systems can be interpreted in terms of different line widths which can be explained to be a consequence of a coupling to the OH proton. Normally, CH–OH couplings are not visible in ^1H-NMR spectra because of fast intermolecular exchange of OH protons. Very strong hydrogen bonding, e.g., with dimethyl sulfoxide as a solvent, reduces the exchange rate to such an extent that the couplings become visible.

Consult SpecTool:
Navigate to the **HNMR Data** page **Aliphatic Alcohols** and consult the pop-ups for **Shift** and **Coupling**, and use the hot word link to access the data page **Alcohols in Dimethyl Sulfoxide.**

In the present case, the intramolecular hydrogen bond is not strong enough to make the coupling obvious but is sufficient to cause line broadening. Vicinal couplings strongly depend on the dihedral angle showing maxima at 0 and 180° and minima at 90°. Owing to the preferred conformation, there is a significant difference between the two HO–CH coupling constants and, therefore, the line widths of the two methylene protons.

The two tiny signals near 1.05 and 0.72 ppm are due to ^{13}C–^{1}H coupling (^{13}C-satellites). Their distance from the chemical shift value of the corresponding protons in absence of ^{13}C-isotopes equals half the direct ^{13}C–^{1}H coupling constant (the isotope effect of ^{13}C on the ^{1}H-chemical shift is negligible in routine analysis). Generally, each signal is accompanied by spinning side bands in addition to the ^{13}C-satellites. The spacing of these side bands is also symmetric about the main signal and equals the spinning frequency of the sample tube or an integral multiple thereof. They are caused by magnetic field inhomogeneities and exhibit in general an intensity of $< 0.5 - 1\%$ of the main band.

Consult SpecTool:
Access the ^{1}H-NMR spectrum in **SpecLib**, measure the separation of these lines and check whether it is compatible with the one-bond ^{13}C-^{1}H coupling constant. You find reference values under **CNMR Data**, **Special**, $^{1}J_{CH}$.

3 Problem 3

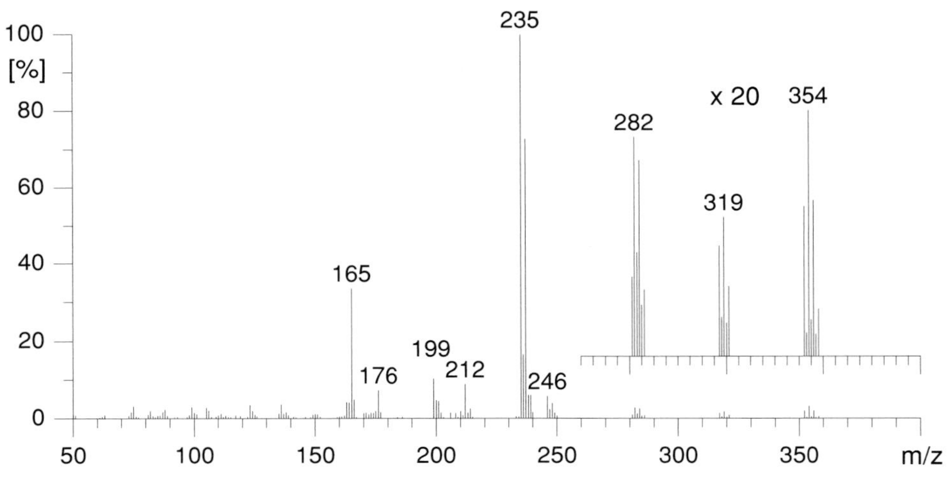

Fig. 3.1: Mass spectrum: EI, 70 eV

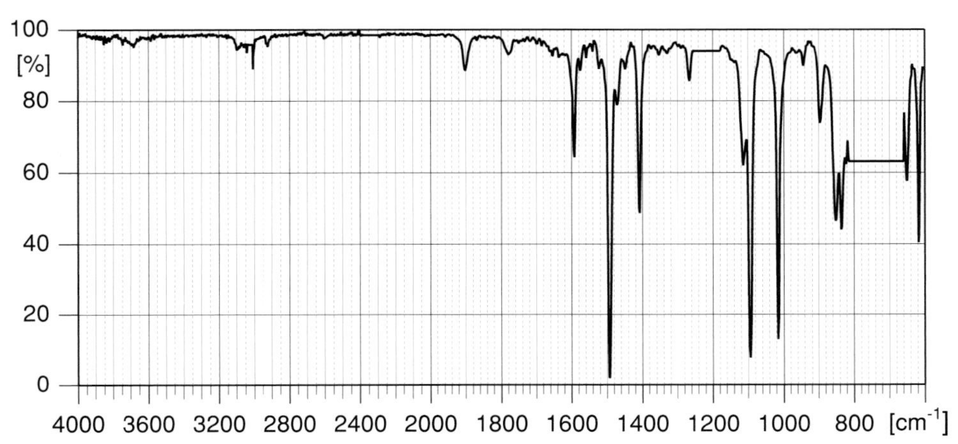

Fig. 3.2: IR spectrum: recorded in $CHCl_3$, cell thickness 0.2 mm

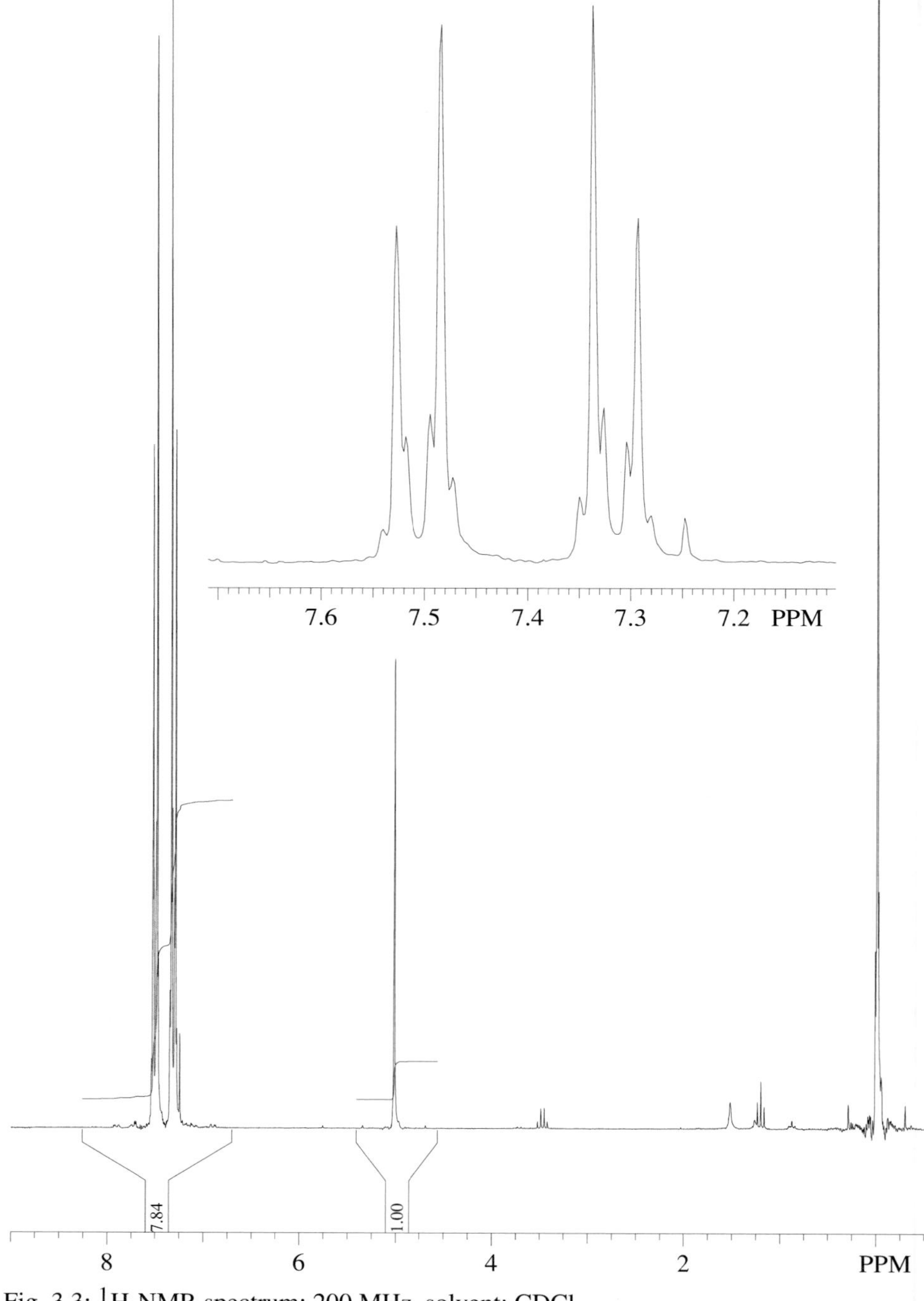

Fig. 3.3: ^{1}H-NMR spectrum: 200 MHz, solvent: $CDCl_3$

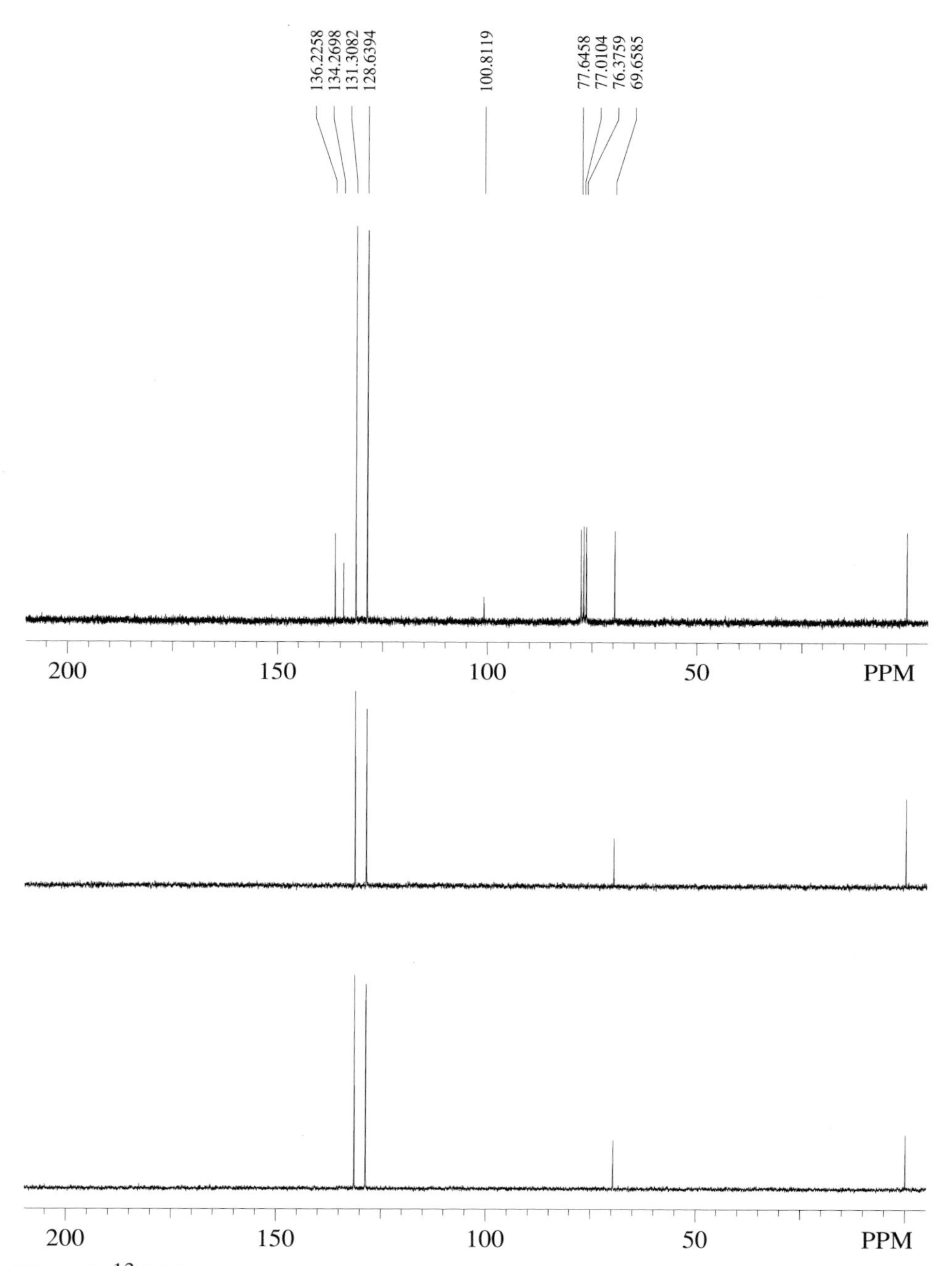

Fig. 3.4: ^{13}C-NMR spectrum: 50 MHz, solvent $CDCl_3$, Top: proton decoupled, middle: DEPT135, bottom DEPT90 ($\tau = 3.45$ ms)

3.1 Elemental Composition and Structural Features

Infrared spectrum: The high frequency range of the spectrum is practically devoid of signals. Weak combination bands at 1900 cm^{-1} and 1780 cm^{-1} and sharp bands at 1600 cm^{-1} and 1500 cm^{-1} indicate an aromatic system.

Proton NMR spectrum: An *AA'BB'* coupling pattern at $\delta = 7.3$ and 7.5 representing eight protons corroborates the aromatic nature and indicates symmetrical substitution. In addition, only a methine proton singlet at $\delta = 5.0$ is observed, strongly deshielded by three substituents.

Carbon-13 NMR spectrum: In the DEPT spectra the presence of a methine proton is borne out by the signal at $\delta = 69.7$. A symmetrically disubstituted aromatic moiety is indicated by two signals for CH and two for C between $\delta = 136.2$ and $\delta = 128.6$. Furthermore, a deshielded quaternary carbon atom is shown by a signal at $\delta = 100.8$.

Mass spectrum: The spectrum is terminated by a peak cluster around m/z 354, with two mass units difference between individual maxima within the group. Such a situation is always indicative of the isotope pattern caused by combination of several chlorine and/or bromine atoms. Intensity distribution within the group, together with a sequence of mass differences of 35, 36 and 70 mass units with adequate changes of intensity pattern prove that five chlorine atoms are present.

Consult SpecTool:
An overview over the simple chlorine and bromine isotope patterns can be found in the **MS Data** section **Halogen Compounds**.

Loss of three of these halogen atoms is involved in formation of a base peak of dominating intensity at m/z 235. Since the mass difference from the molecular ion (by definition m/z 352, if the last group of peaks represents the molecular mass) is 117, the lost neutral entity must be a trichloromethyl group with an easily fragmenting bond. The ^{13}C-isotope peak at m/z 236 has 15% relative intensity, which means that up to 13 carbon atoms can be present in m/z 235. This fits beautifully to two aromatic rings, which are necessary anyway to account for eight protons in the proton NMR spectrum and which must be attached to the methine carbon atom. Loss of two chlorine atoms from the base peak produces the next most prominent fragment at m/z 165, which fits the expected elemental composition $C_{13}H_9$ arising from a molecular formula $C_{14}H_9Cl_5$ with eight double bond equivalents.

3.2 Structural Assembly

Two *para* chlorophenyl groups, one trichloromethyl group, and a methine carbon atom can only be arranged in one way (to 1,1-bis(4-chlorophenyl)-2,2,2,-trichloroethane, DDT):

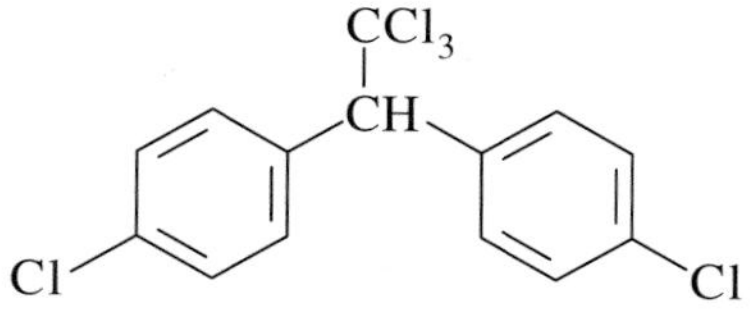

3.3 Comments

3.3.1 Mass Spectrum

Except for the base peak, most fragments arise by consecutive losses of Cl, HCl, and Cl_2.

Check it in SpecTool:
An external program allows for the calculation of the **Isotope Pattern** of an arbitrary complex species. Use it to predict the isotope patterns for the peak clusters from m/z 235 upwards and compare with the experimental data. The program is available in **MS Tools**. (Use the system menu to exit the **Isotope Pattern** program).

3.3.2 Infrared Spectrum

The type of aromatic substitution could be deduced from the presence of only two combination bands at 1900 and 1780 cm^{-1} and their intensity ratio. The prominent bands at 1095 and 1015 cm^{-1} result by combination of C–Cl and C–C stretching vibrations, whereby the one at 1095 cm^{-1} is characteristic for a *para* substituted chlorophenyl group.

Check it in SpecTool:
Besides the aromatic C-Cl, there are also bands for the aliphatic C-Cl stretching vibration. Data for both can be found in the **Halogen Compounds** section of **IR Data**. Try to assign those bands not masked by solvent absorption.

3.3.3 Proton and Carbon-13 NMR Spectra

The additivity rules for the estimation of the chemical shifts of aliphatic carbon atoms often lead to unreliable results in polyhalogenated compounds and also in cases with extreme chemical shift values. In the present case, however, the predicted values are reasonably close to the experimental ones.

Check it in SpecTool:
Predict the chemical shifts. After starting **1H Shift Estimation** on the **HNMR Tools** page, or **C13 Shift Estimation** on the **C13 Tools** page, the demo version of **ChemWindow® III** is launched, where you can now draw the target structure (if you are not familiar with **ChemWindow® III**, you find an introduction in Chapter 19). The estimation tools require every node to be specified by drawing, therefore do not draw hydrogen atoms and use letters only for heteroatoms. After drawing the molecule, select it and copy it into the clipboard. Exit **ChemWindow® III** (chose Exit from the File menu) and start the calculation with the **Estimate** button. The results are written in a new **ChemWindow® III** window. Please note that its demo version does not allow you to save or to print the results and that the use of the shift estimation programs is limited to the examples covered by this book.

4 Problem 4

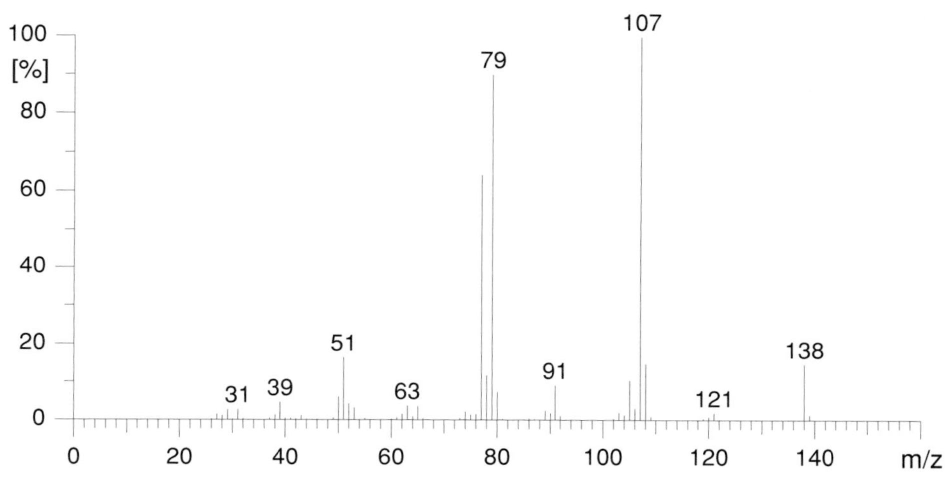

Fig. 4.1: Mass spectrum: EI, 70 eV

Fig. 4.2: IR spectrum: recorded in $CHCl_3$, cell thickness 0.2 mm

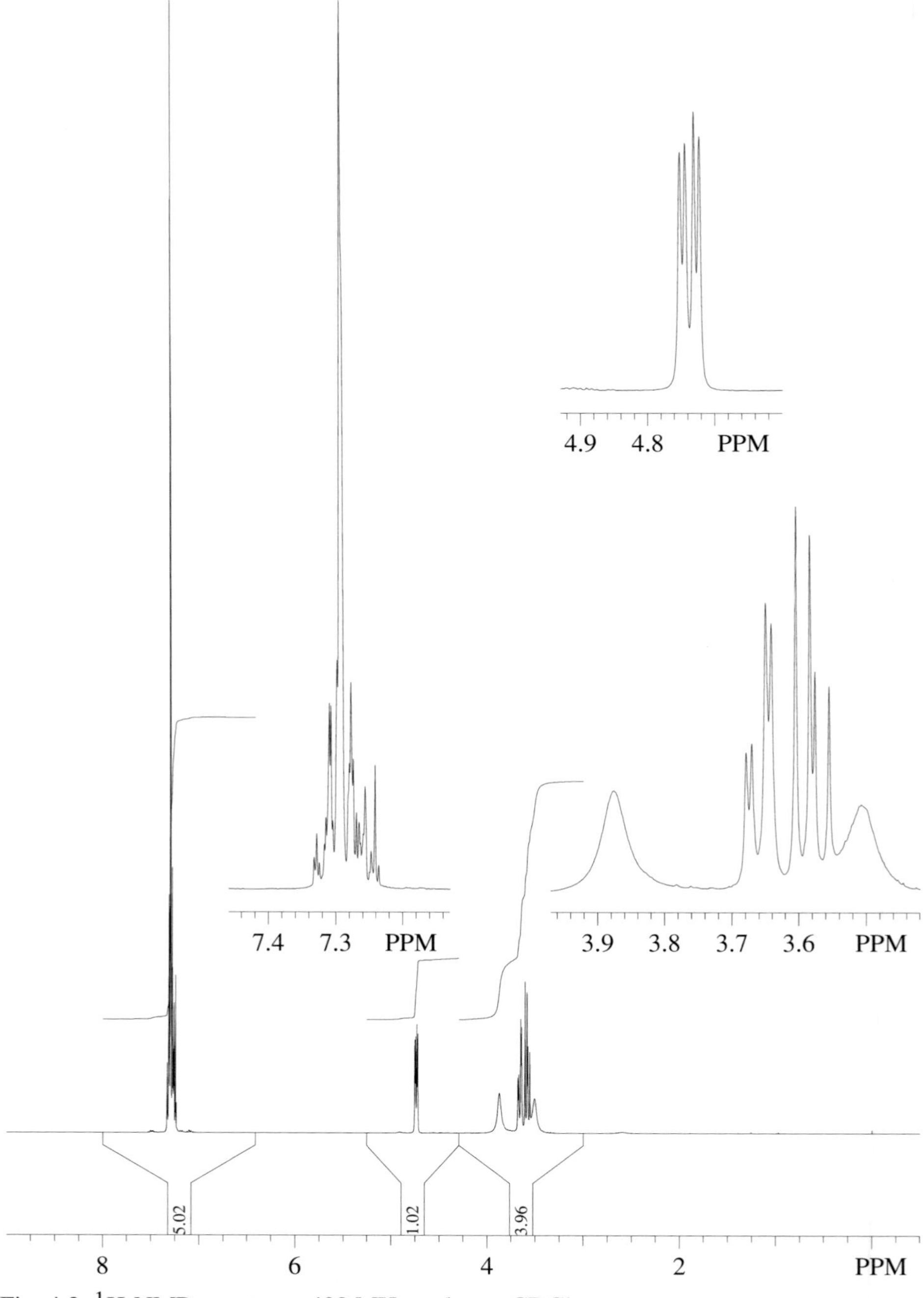

Fig. 4.3: ^{1}H-NMR spectrum: 400 MHz, solvent: $CDCl_3$

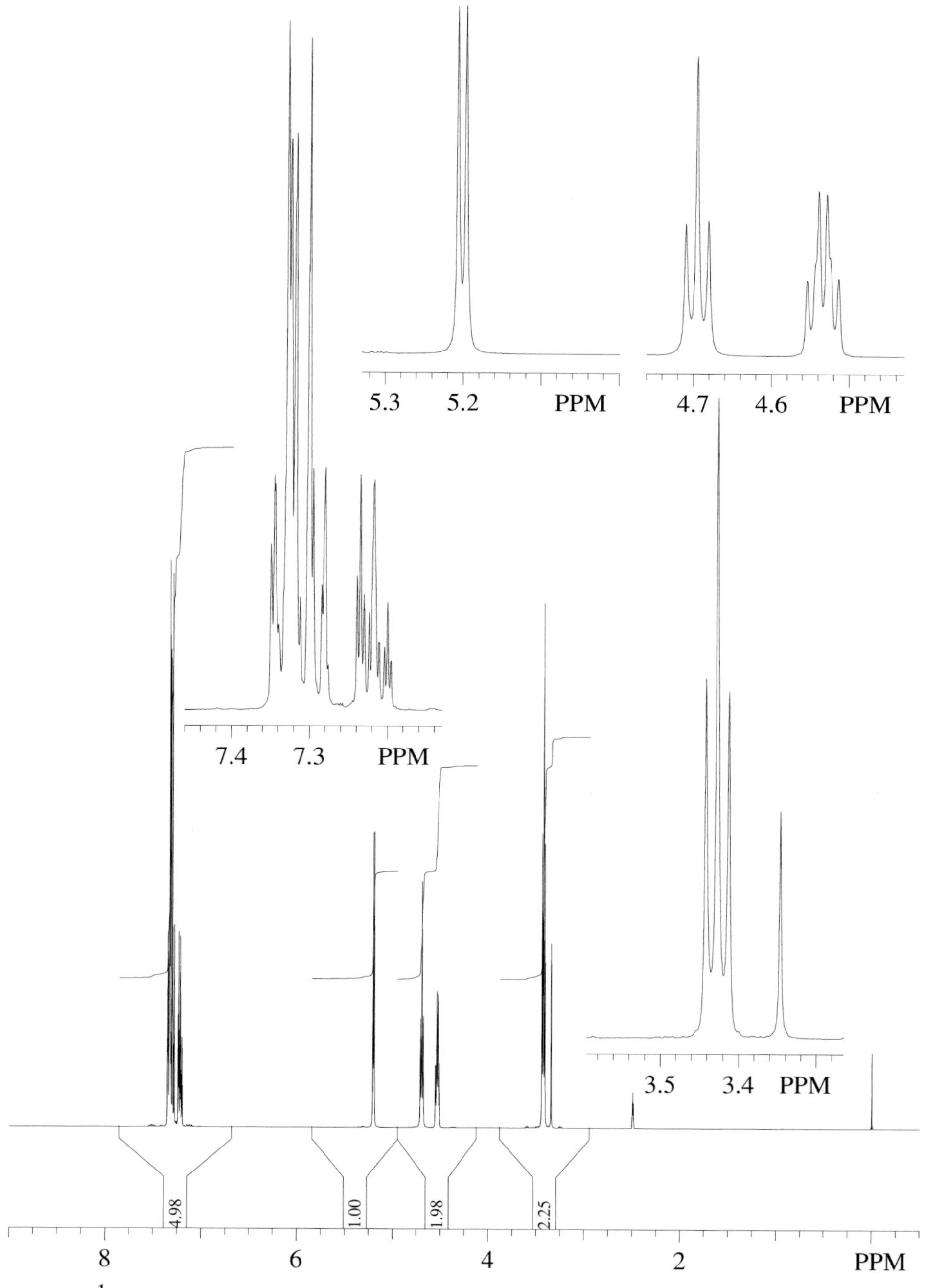

Fig. 4.4: ^{1}H-NMR spectrum: 400 MHz, solvent: dimethyl sulfoxide (DMSO)

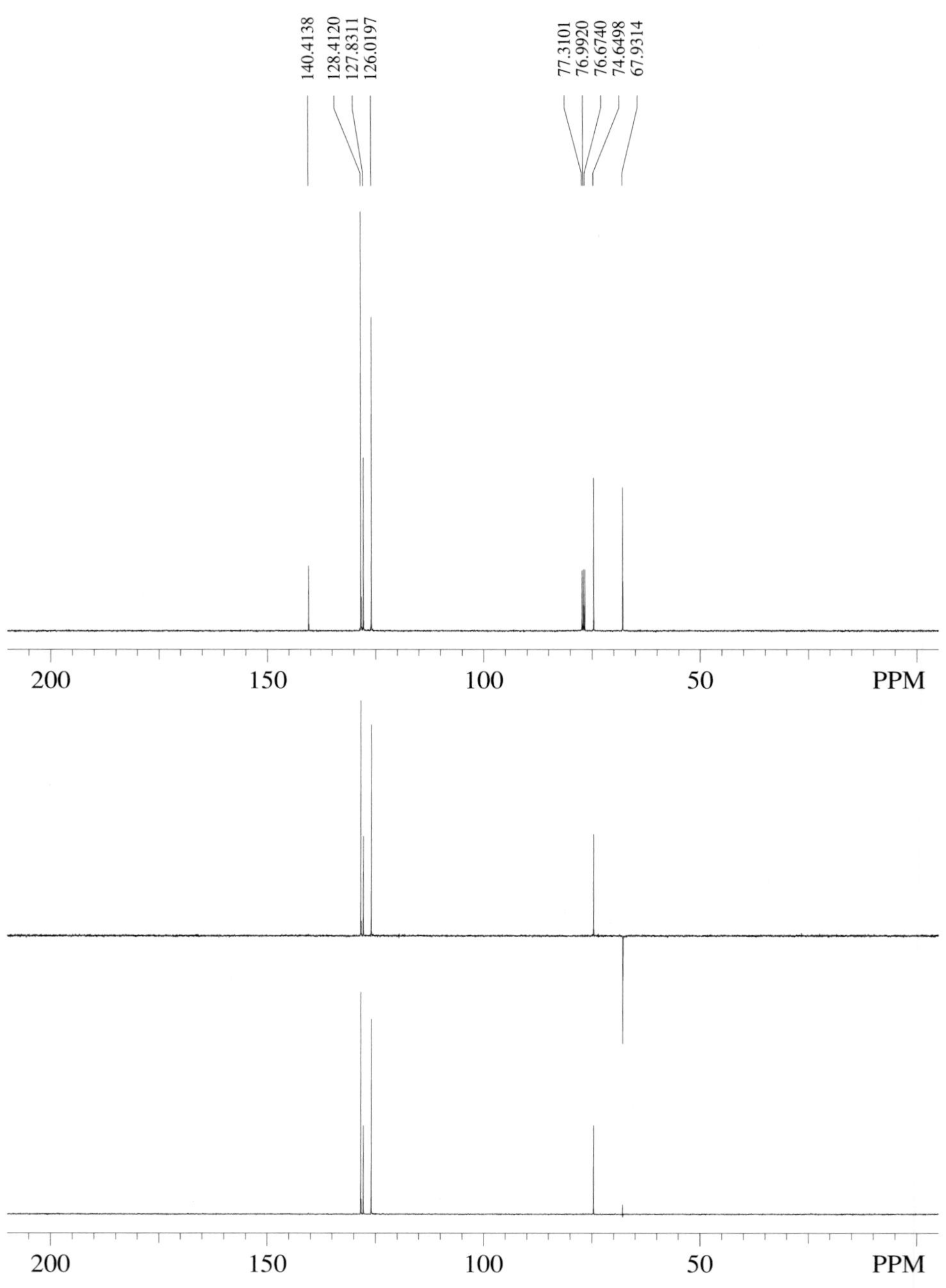

Fig. 4.5: ^{13}C-NMR spectrum: 200 MHz, solvent $CDCl_3$, Top: proton decoupled, middle: DEPT135, bottom DEPT90 ($\tau = 3.6$ ms)

4.1 Elemental Composition and Structural Features

The infrared spectrum gives the following structural information: OH present (3600 cm^{-1}: free OH stretching vibration, 3425 cm^{-1}: associated OH st.), phenyl ring, probably monosubstituted (2000 to 1600 cm^{-1}: combination vibrations, 1600 and 1500 cm^{-1}: skeletal vibrations), no carbonyl groups.

Consult SpecTool:
Compare the IR combination band patterns with the reference data shown under **IR Data**, **Aromatic Compounds**, **Overtone and Combination bands**.

The mass spectrum ends with m/z 138. The differences to the other peaks being chemically reasonable we tentatively assume a molecular mass of 138. The benzene ring is confirmed (m/z 39, 51, 63-65, 77). We expect an sp^3 hybridized carbon atom as a substituent (m/z 91).

Consult SpecTool:
If you are still on the **IR** page accessed above, just change the method to **MS** (by using the **Menu Switch** or **Palette SWT**), and select **Aromatic Hydrocarbons** to check the ions series. Otherwise access the page from **Top**, **Aromatic Compounds**, **Aromatic Hydrocarbons**. Then switch to **Tools**, select **Homologous Mass Series** and compare the fragments expected for various compound classes.

For a monosubstituted phenyl ring we expect four signals in the carbon-13 NMR spectrum, three of them for CH and one for C as inferred from the DEPT spectra. This is just what we find in the aromatic region. In addition, another CH and a CH_2 group are present. Thus we expect 5 + 1 + 2 = 8 protons in the proton NMR spectrum. In addition, we shall find the proton(s) on the oxygen atom(s). Integration of the proton NMR spectrum taken in deuterochloroform gives (from left to right) a proton ratio of 5 : 1 : (1: 3), thus indicating two hydroxyl groups. Integration of the spectrum recorded in DMSO (dimethyl sulfoxide) also results a total of 10 protons (5 : 1 : 1 : 1 : 2) but very different chemical shifts and coupling constants are observed (see Comments).

Consult SpecTool:
Identify the signals originating from the solvent. Use information on the **Common Solvents** page accessible from the page **Special Data** in **HNMR Data**.

4.2 Structural Assembly

We may now summarize the results obtained so far. We have a monosubstituted phenyl ring, one methylene group, one methine group, and two hydroxyls. This sums up to 138 Daltons, indicating that all constitutional elements have been identified. There is only one possibility of connecting them in a chemically meaningful way, namely:

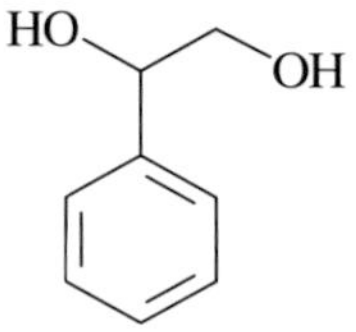

4.3 Comments

4.3.1 Mass Spectrum

Base peak formation results from benzylic cleavage as would be predicted, subsequent decarbonylation (m/z 107 → 79) is typical for deprotonated benzyl alcohol cations. Water elimination from the molecular ion probably results in styrene oxide, which is known to lose CHO (mass 29) after rearrangement to the isomeric aldehyde. This reaction sequence gives rise to m/z 91, as described in the simplified formalism:

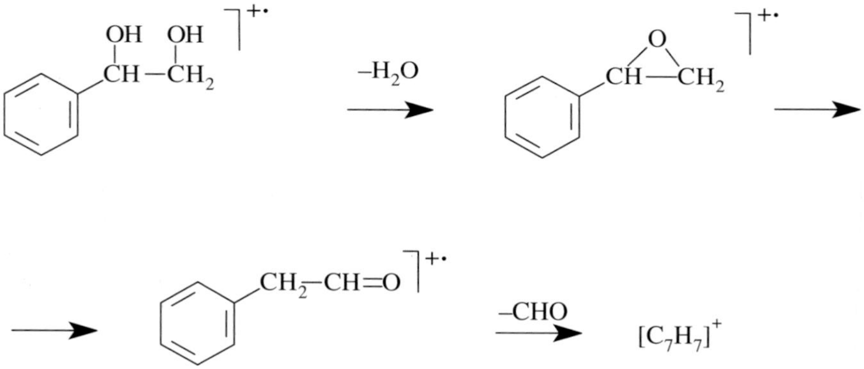

$$[C_6H_5-CH(OH)-CH_2(OH)]^{+\cdot} \xrightarrow{-H_2O} [C_6H_5-CH(O)CH_2]^{+\cdot} \longrightarrow [C_6H_5-CH_2-CH{=}O]^{+\cdot} \xrightarrow{-CHO} [C_7H_7]^+$$

Check it in SpecTool:
Compare the proposed mechanism with the fragmentation path of benzyl alcohol shown under **Alcohols** in the **MS Data** section.

4.3.2 Infrared Spectrum

Prominent skeletal vibration frequencies for benzene rings are observed near 1600 cm^{-1}, from around 1500 cm^{-1} to 1450 cm^{-1} and near 700 cm^{-1}. The respective vibrations may be described as quadrant stretching, semicircle stretching, and sextant bending. The two former vibrations consist of two components each, which can often be resolved in the spectrum.

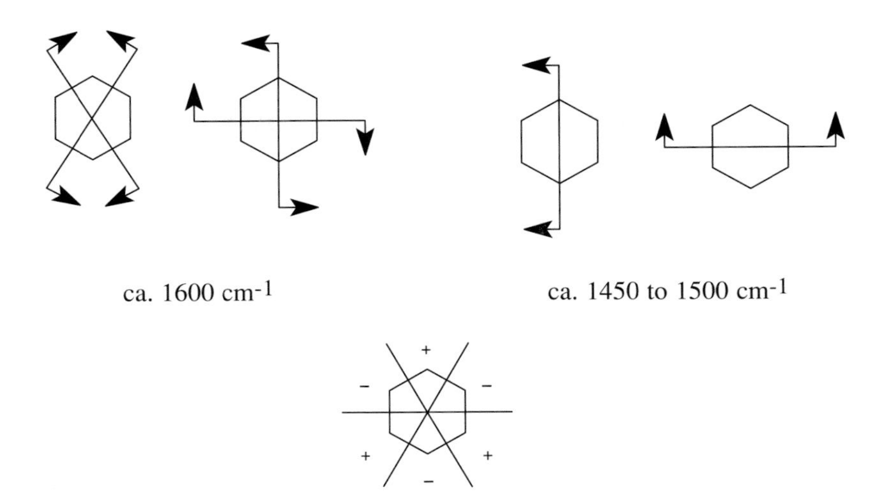

ca. 1600 cm^{-1} ca. 1450 to 1500 cm^{-1}

ca. 700 cm^{-1}

The 1600 cm^{-1} vibrations are absent in compounds having a centre of symmetry in the centre of the ring. In monosubstituted benzene rings the intensity is high for substituents that are either strong electron acceptors or strong electron donors. For weakly interacting substituents the intensity is low. In *para* disubstituted compounds the intensity is proportional to the difference in electronic effects of the two substituents, in *meta* disubstituted compounds it is proportional to their sum. *Ortho* disubstituted derivatives are intermediate. The 1500 cm^{-1} bands are always present, although sometimes rather weak, if e.g., the substituents are carbonyl groups.

The band at 700 cm^{-1} is absent for compounds exhibiting a centre of symmetry in the centre of the ring, and for *ortho* disubstituted compounds, if the substituents are identical. Even with non identical substituents the intensities are low in *ortho* and *para* disubstituted compounds. High intensity absorptions are generally observed in mono-, *meta*-di-, and symmetrical trisubstituted benzene compounds.

The intensities observed for the respective bands in the infrared spectrum of the present compound fit nicely into this scheme.

> **Consult SpecTool:**
> These rules are summarized on the corresponding **IR Data** page which also gives access to specific reference data.

In benzene systems a series of low intensity bands is observed between 2000 cm^{-1} and 1660 cm^{-1}. These combination bands arise from various interactions of C–H deformation vibrations and skeletal vibrations of the ring. They show a characteristic pattern that depends on the number of vicinal hydrogen atoms on the benzene ring and are thus indicative of the substitution type. The exact position of the bands is rather variable. However, the overall pattern is reasonably constant. It is thus advisable to consult suitable reference spectra when assigning substitution types rather than numerical lists of band positions. In standard spectra the intensity of these combination bands is generally too low to be of practical value. An exception are monosubstituted and *para* disubstituted benzene rings. The former exhibit three equally spaced absorptions of discernible intensity, whereas the latter are characterised by one relatively strong absorption band. The correlation between the appearance of the spectrum in this range and the substituent pattern is quite reliable as long as there are no substituents which interact strongly with the π-system of the ring. With such substituents (e.g., C=O, NO_2, C=C, etc.), the correlation becomes notoriously unreliable and misleading.

Check it in SpecTool:
Click at the IR combination band patterns in **IR Data**, **Aromatic Compounds**, **Overtone and Combination bands** to get access and to reference data.

4.3.3 Proton NMR Spectrum

The vicinal coupling between the hydroxylic proton and the proton on the neighbouring carbon atom generally does not lead to observable splittings in the proton NMR spectrum, because of the fast intermolecular exchange of the hydroxyl protons. Traces of acids (generally present in deuterochloroform) or bases catalyze this exchange reaction. For very pure alcohols in absolutely acid- and base-free solvents, the exchange rate is often slow enough to make these couplings observable. In the present case it is slow relative to the chemical shift difference of the two hydroxyl protons (0.37 ppm i.e., 148 Hz) but fast relative to the vicinal coupling constants of ~7 Hz. Therefore, two distinct signals are observed, which are broadened due to the exchange, but no coupling to the vicinal CH_2 and CH is seen in the spectrum. The mean lifetime τ [s] of the species is accordingly:

$$\frac{1}{148} < \tau < \frac{1}{7}$$

In dimethyl sulfoxide as solvent, the couplings are generally observed if the compound does not exhibit acidic or basic groups and if the sample does not contain such impurities. Because of the slow exchange rates, distinct chemical shifts are generally observed for the non-equivalent OH protons (including water at $\delta = 3.4$). The number of lines in an OH multiplet indicates directly whether the alcohol is primary, secondary, or tertiary. The reason for this unique effect of dimethyl sulfoxide is the strong hydrogen bonding between hydroxyl groups and the highly polar sulfoxy group. Since the formation of hydrogen bonds leads to a deshielding of the protons involved, the chemical shifts of the hydroxyl groups in aliphatic alcohols lies in the region of 4 to 6 ppm (in dilute chloroform solutions typically 1 to 2 ppm). In our case, the doublet at $\delta = 5.2$ and the triplet at $\delta = 4.7$ in the proton NMR spectrum recorded in dimethyl sulfoxide solution, indicate that both a primary and a secondary alcoholic group are present.

In $CDCl_3$ as solvent, the two diastereotopic methylene protons have different chemical shifts ($\delta = 3.56$ and 3.66) and couple differently to the vicinal methine proton (J = ca. 8.9 and 3.3 Hz, respectively). This information could be the base of an educated guess about the prevailing conformation. The staggered conformations given below are considered:

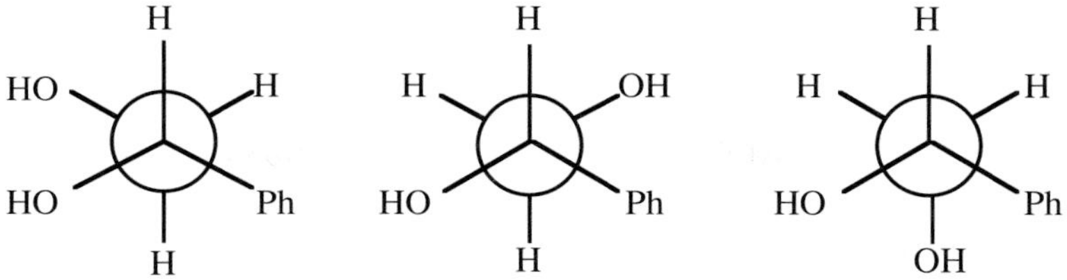

I **II** **III**

For a torsion angle of 180° one expects a large coupling constant in the order of 10 Hz, whereas for 60° the coupling constant is small, ca. 4 Hz. This rules out conformation **III** since two small vicinal coupling constants are expected for it. A detailed calculation taking into account the influence of the substituents gives a slightly better fit for conformation **I**, which is also the one expected because of the possibility of an intramolecular hydrogen bond.

Check it in SpecTool:
From the **TOP** page select **HNMR Tools**, and then **Vicinal coupling**. The Tool allows the estimation of vicinal coupling constants as a function of the torsional angle and the substituents. Verify the above statements. Use C-substituents for the phenyl group.

In dimethyl sulfoxide the spectrum is markedly different: The two methylene protons have become isochronous and, the coupling constant to the CH and the OH being accidentally the same, the signal is a triplet. The methine proton shows four lines, a closer inspection reveals a doublet of a triplet. A comparison with the splitting of the corresponding OH proton at 5.2 ppm shows that the smaller coupling is due to the CH-OH and the two identical larger ones of ca. 5 Hz are CH–CH_2 couplings. The conformation(s) in dimethyl sulfoxide must, therefore, be different from that observed in chloroform. This is a consequence of the hydrogen bonds to the solvent. The value of the vicinal coupling constants is with 5 Hz between the extreme values found above in $CDCl_3$. However, it cannot be decided whether there is one single preferred conformation or a dynamic conformation equilibrium.

4.3.4 Carbon-13 NMR Spectrum

Generally, the line intensities in the usual carbon-13 NMR spectra are not proportional to the number of carbon atoms which are represented by the signal (cf. Chapter 19.3). However, for carbon atoms directly attached to hydrogen(s) similar relaxation times and consequently similar line intensities are to be expected, even if the spectrum is recorded under conditions of partial saturation (provided that the C–H vectors exhibit comparable mobilities). Therefore, the signal at δ = 127.8 can be assigned on the basis of its lower intensity to the aromatic carbon atom in *para* position relative to the substituent.

5 Problem 5

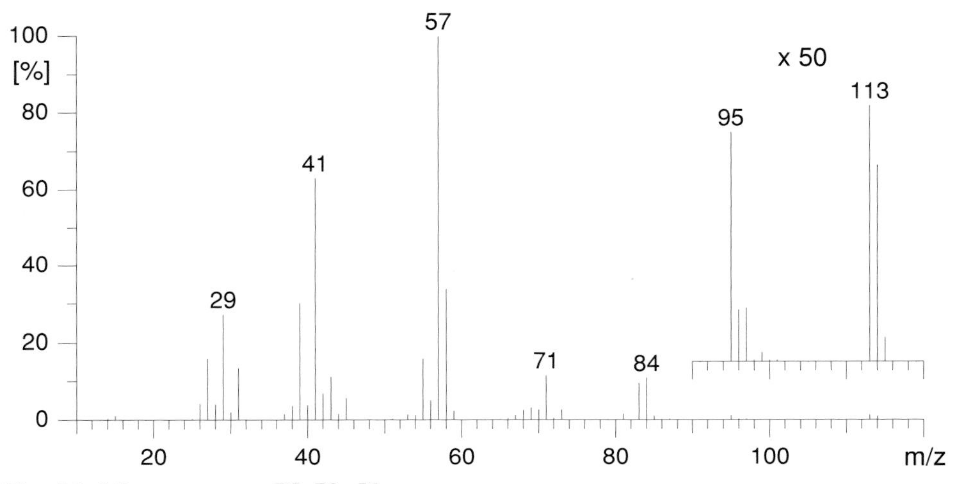

Fig. 5.1: Mass spectrum: EI, 70 eV

Fig. 5.2: IR spectrum: recorded in $CHCl_3$, cell thickness 0.2 mm

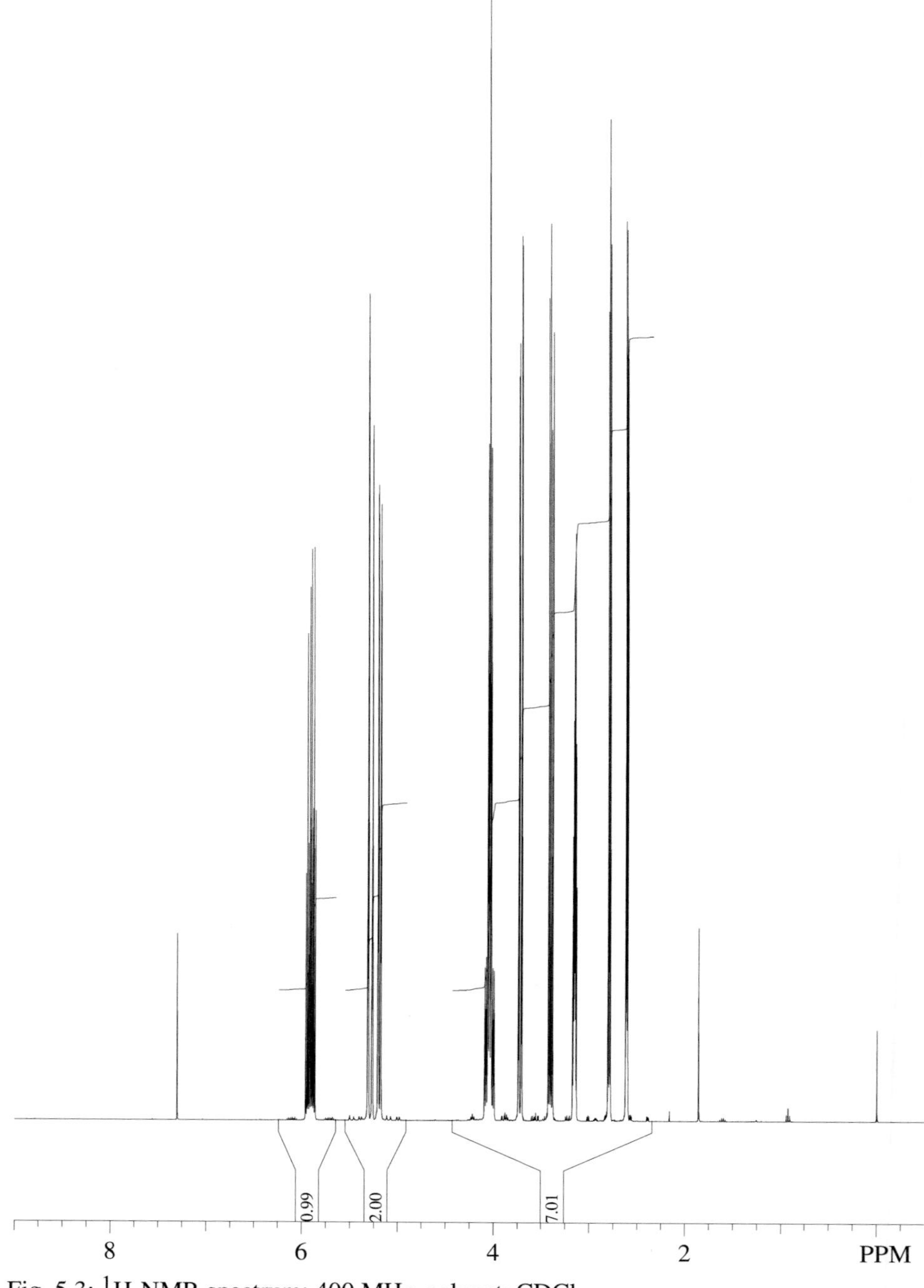

Fig. 5.3: ^{1}H-NMR spectrum: 400 MHz, solvent: $CDCl_3$

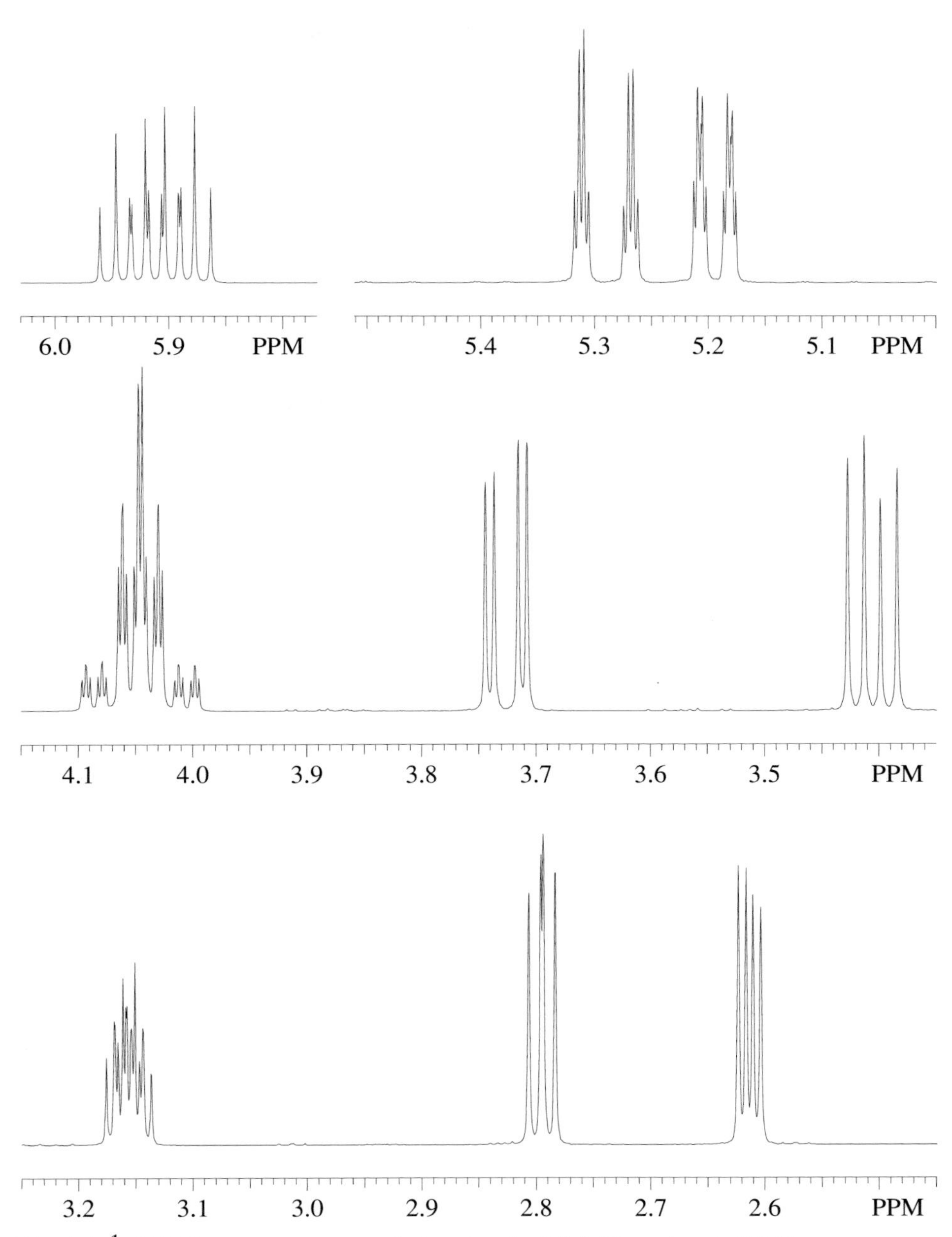

Fig. 5.4: ^{1}H-NMR spectrum (expanded regions): 400 MHz, solvent: $CDCl_3$

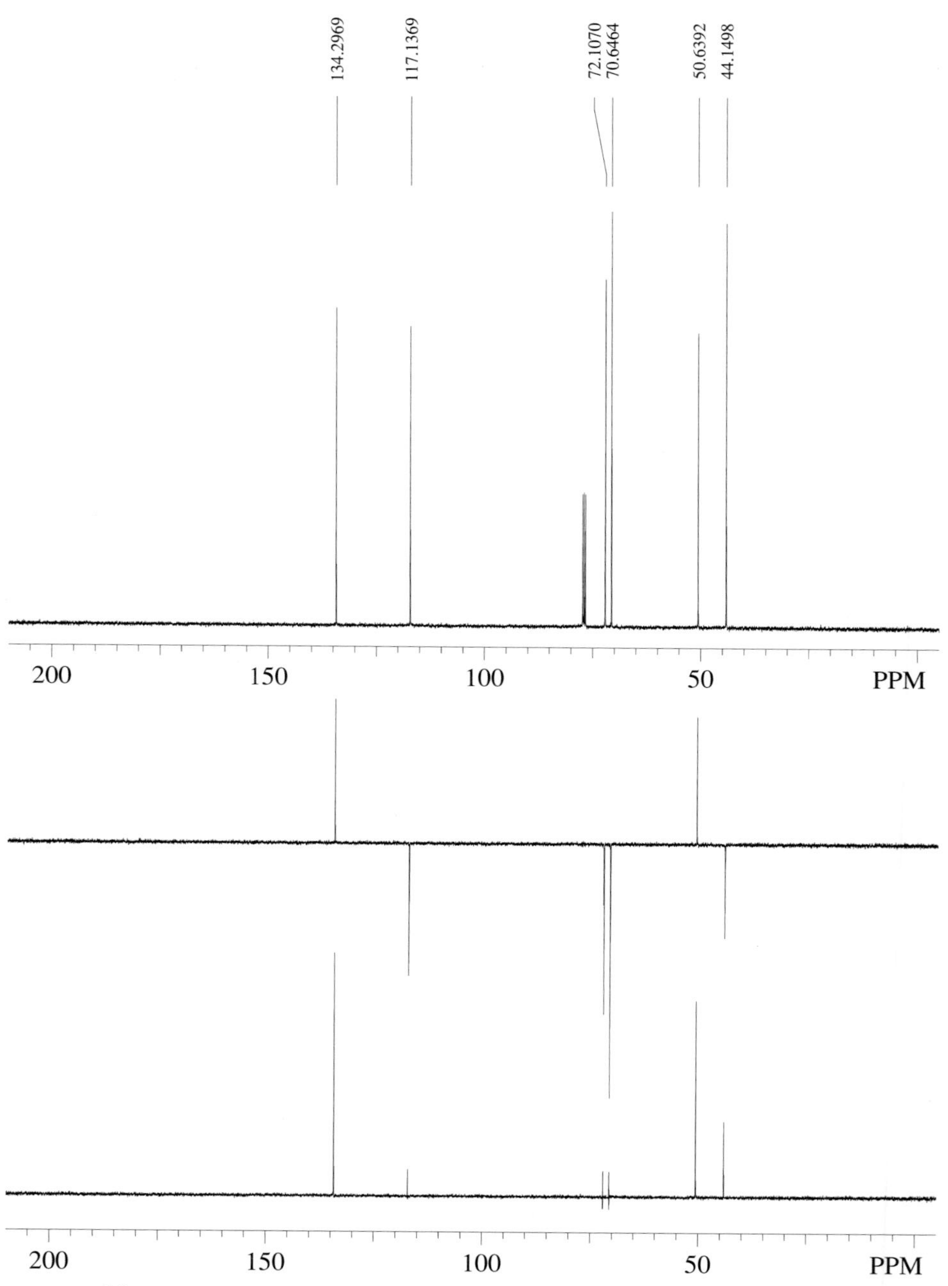

Fig. 5.5: ^{13}C-NMR spectrum: 100 MHz, solvent $CDCl_3$, Top: proton decoupled, middle: DEPT135, bottom DEPT90 ($\tau = 3.6$ ms)

5.1 Elemental Composition and Structural Features

According to the carbon-13 NMR DEPT spectra, the six nonisochronous carbon atoms correspond to four methylene and two methine groups. The signals at $\delta = 117.1$ and $\delta = 134.3$ clearly indicate an olefinic double bond which must be monosubstituted since it carries three protons as shown by the DEPT spectra.

The integration in the proton NMR spectrum leads to the following relative intensities: 1 : 2 : 2 : 1 : 1 : 1 : 1 : 1, 10 protons in total. As the same number of protons has been observed in the carbon-13 NMR spectrum all protons must be bonded to carbon atoms.

In the infrared spectrum we find a confirmation for the vinylic group: C–H stretching vibration at 3060 cm^{-1} C=C stretching vibration at 1640 cm^{-1} and C–H out-of-plane deformation vibrations below 1000 $cm^{-1.}$ An intense band at 1090 cm^{-1} is recognized (it could possibly be a C–O stretching vibration of an ether).

Consult SpecTool:
Compare with the prototype spectra given in the **IR Data** section: **Nonconjugated Alkenes**, **Alkanes**, and **Ether.** Together they cover virtually all bands seen in the spectrum of the compound at hand.

According to the mass spectrum, the molecular mass could be 113 or 114. In the former case m/z 114 is too strong for a ^{13}C-isotope peak so it must predominantly be the protonated molecule ion. If, on the other hand, m/z 114 corresponds to the molecular ion, m/z 113 is formed by the loss of a hydrogen radical. The first alternative can be excluded because we have an even number of hydrogen atoms and there are no indications for halogens. An odd nominal molecular mass implies an odd number of nitrogen atoms in the molecule which, in turn, implies an odd number of odd-valence elements.

The molecule fragments easily in two halves of equal mass (m/z 57). The elements found so far (C_6H_{10}) correspond to a mass of 82 units leaving 32 mass units to complete the molecule. This is most readily accomplished by adding two oxygen atoms or one sulfur atom. The mass spectrum does not contain any indication for sulfur:

Consult SpecTool:
Select the page **MS-Tools**, **Indicators for Heteroatoms**, and look for indication of S.

Furthermore, the carbon-13 chemical shifts of around $\delta = 70$ for the aliphatic methylene groups demand the presence of oxygen. The elemental composition thus becomes $C_6H_{10}O_2$ corresponding to two double bond equivalents. There is only one C=C and no C=O double bond (see the information in the infrared and carbon-13 NMR spectra) so that the molecule contains one ring.

5.2 Structural Assembly

The vinyl group is not bonded to an oxygen atom as indicated by the chemical shifts in both NMR spectra since for a CH_2=CH–O– group proton shift values of δ = 6.4, 4.0 and 3.9 and carbon-13 chemical shift values of δ = 150 and δ = 85 can be expected:

> **Consult SpecTool:**
> Look in the tables **Monosubstituted Ethylenes** in the C13-NMR-Data and H1-NMR Data sections. Note that you can directly jump between these two tables by selecting the appropriate method in the **Switch Menu** or **Switch Palette.** Note also the ^{1}H-NMR coupling constants for later use.

The two signals around 5.2 and 5.3 ppm can be assigned to the two geminal protons since they show the required *cis* and a *trans* coupling of ca. 10 and 18 Hz, respectively (vicinal couplings with the proton at 5.9 ppm). One would expect two doublets, if no other couplings were present. However, the signals can be further split by a small geminal coupling (0 - 4 Hz) and possibly by allylic couplings. In the present case, the lines of both protons are split into quartets. This can be rationalized by the presence of a CH_2 substituent attached to the vinyl group which gives rise to an allylic coupling of approximately 2 Hz, being by chance equal to the geminal coupling constant. The presence of a CH_2 substituent is corroborated by the splitting pattern of the third vinylic proton at 5.9 ppm which can be resolved into four triplets. The two protons of the allylic CH_2 group should also show the allylic couplings. The signal group at 4.05 ppm fulfills this requirement. The signal pattern, however, is rather complicated at the first glance indicating that the two methylene protons are anisochronous. As a consequence, the geminal coupling shows up and leads to a higher order spectrum which, however, can be understood by first order rules (see Comments). The chemical shift value of δ = 4.05 shows that the allylic methylene group is bonded to another deshielding substituent. The only possibility for such a substituent with the elemental composition found above is an oxygen atom. Thus the group CH_2=CH–CH_2–O– (mass 57) constitutes one half of the molecule. The following possibilities can be considered for the other half with the same elemental composition and one ring:

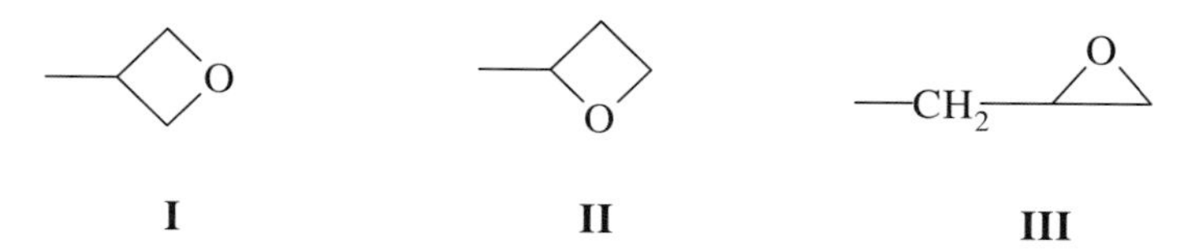

Substructure **I** can be excluded because the data preclude such a symmetry. In **II** we would have an O–CH–O group. Such groups have proton chemical shifts of around δ = 6 and carbon-13 chemical shifts of δ = 90 to 100. Since no resonances in these ranges occur in the NMR spectra, the solution **II** can be excluded with certainty. The remaining structure is thus:

5.3 Comments

5.3.1 Mass Spectrum

The mass spectrum nicely reflects the general rule that upon cleavage of bonds to hetero atoms the positive charge stays preferentially with the carbon atom due to effects of electronegativity. The two most intense peaks at m/z 57 and 41 are the result of fragmentation at the ether oxygen atom, m/z 58 is due to allyl alcohol cation obtained by hydrogen rearrangement and presumably is the precursor of m/z 31. The observation of fragments at m/z 31, 45, 58, and 73 could be used as direct evidence to indicate the presence of oxygen.

Check it in SpecTool:
For fragmentation rules access **MS-Data - Ether - Aliphatic Ethers** and select the hot-word link to the fragmentation schemes.

5.3.2 Infrared Spectrum

C–H stretching frequencies between 3100 cm^{-1} and 3000 cm^{-1} are generally observed only for hydrogen atoms bonded either to an sp^2-hybridized carbon atom or to a carbon bearing halogen atom(s). In addition, three-membered rings also exhibit C–H stretching frequencies in this range. For aromatic moieties the respective absorption bands are generally quite weak and the frequencies are near the lower limit. Medium intensity bands at the upper frequency limit are observed for terminal methylene groups, for methylene groups in three-membered rings, and sometimes for hydrogen atoms in five-membered heteroaromatic systems. In the present case, the absorption at 3080 cm^{-1} is assigned to the methylene group in the oxirane ring and to the terminal methylene of the vinyl group.

Check it in SpecTool:
Check the respective pages in the **IR-Data** section.

5.3.3 Proton NMR Spectrum

Since the molecule is chiral, the methylene protons are diastereotopic and thus in principle anisochronous. If the chemical shift differences are small relative to the coupling constants, higher order spectra will result. These can often be understood assuming first order splittings. However, borderline cases between first and higher order spectra often contain pitfalls so that the calculation of the spin system is recommended as a control. The same comment holds true for the following remarks: Vinyl groups often lead to *ABC* type spectra which are close to *ABX* cases. The *X*-part of an *ABX* system generally consists of 4 lines (in some cases the theoretically possible six lines are observed) positioned symmetrically around the centre (chemical shift). The symmetry-equivalent lines have the same intensities. In borderline cases between *ABX* and *ABC* systems, there are still four lines which are symmetrical with respect to the position but no longer symmetrical with respect to the intensities. It is important to keep in mind that

ABX systems lead to higher order spectra and the line spacings in the *X*-part are not necessarily equal to the coupling constants.

The basic pattern of the signal at 4.05 ppm, but not necessarily the coupling constants, can be understood as the overlap of two signals, each split into doublet (geminal coupling) × doublet (vicinal coupling) × triplet (two allylic couplings with the same coupling constant). In the center of the pattern two lines overlap, i.e., two triplets mimick a quartet. The *cisoid* and *transoid* allylic coupling constants do not necessarily have the same value. Here they are almost equal.

Check it in SpecTool:
Possibilities and limits of the quantitative usage of allylic coupling constants are summarized in the corresponding **HNMR Data** - **Alkenes** section.

5.3.4 Carbon-13 NMR Spectrum

The assignment of the olefinic carbon atoms is trivial on the basis of their chemical shift values and their signes in the DEPT135 spectrum.

In epoxides, both protons and carbon atoms are more strongly shielded than the corresponding atoms in other cyclic or acyclic ethers. The signals at δ = 44.1 (methylene) and δ = 50.6 (methine) can thus be assigned to the epoxide group. The assignments of the carbon-13 signals for the two remaining methylene groups is not possible on the basis of the data presented here. It would, however, be a trivial task using two dimensional $^{13}C{-}^{1}H$ correlation spectroscopy since the corresponding protons are assigned (allylic methylene: δ = ca. 4.04 and 4.06, other aliphatic methylene: ca. δ = 3.72 and δ = 3.41).

Multipulse techniques for identifying the number of protons attached directly to a carbon atom assume one fixed value for all $^{13}C{-}^{1}H$ one-bond coupling constants. If the real value deviates considerably, additional, generally small, signals may appear (cf. DEPT90 spectrum of this example).

Consult SpecTool:
Consult the table of representative one bond $^{13}C{-}^{1}H$ coupling constants in the **Special** data section of **CNMR Data**. Additionally, there is a **CNMR Tool** for the $^{13}C{-}^{1}H$ coupling estimation for sp^3 hybridized carbon atoms.

6 Problem 6

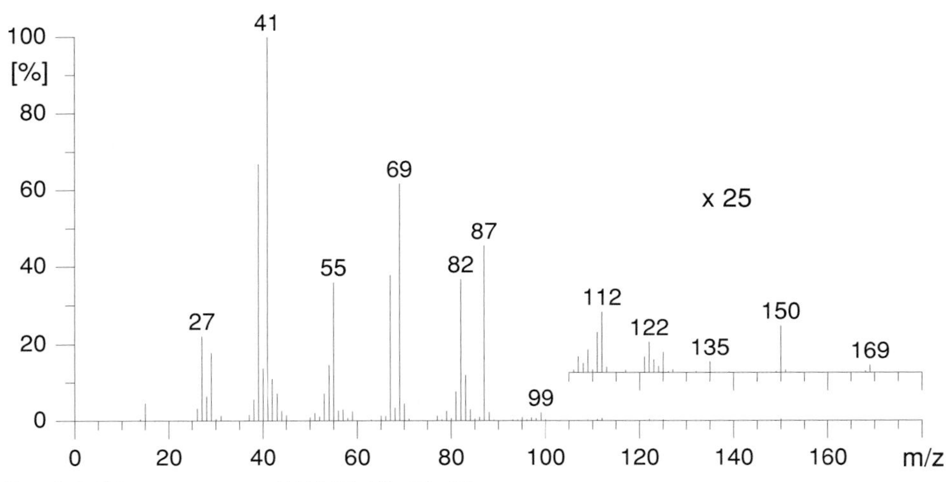

Fig. 6.1: Mass spectrum: GC/MS, EI, 70 eV

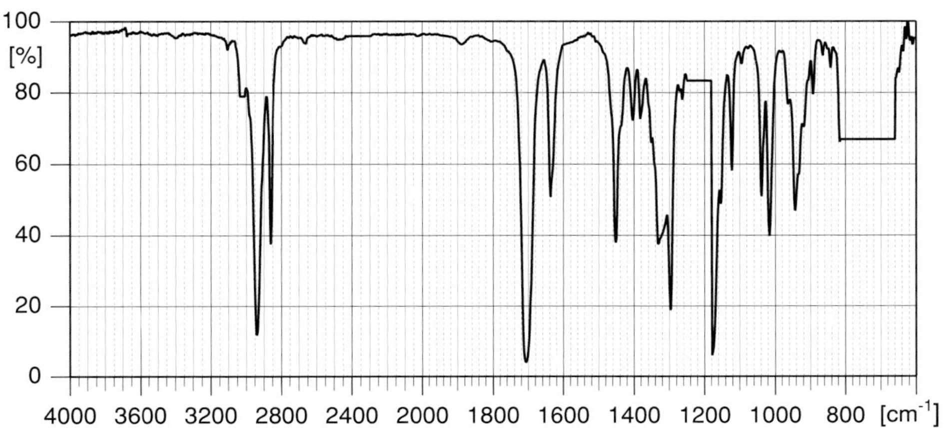

Fig. 6.2: IR spectrum: recorded in $CHCl_3$, cell thickness 0.2 mm

UV spectrum (solvent: ethanol): λ_{max}: 208 nm, log ε: 3.9

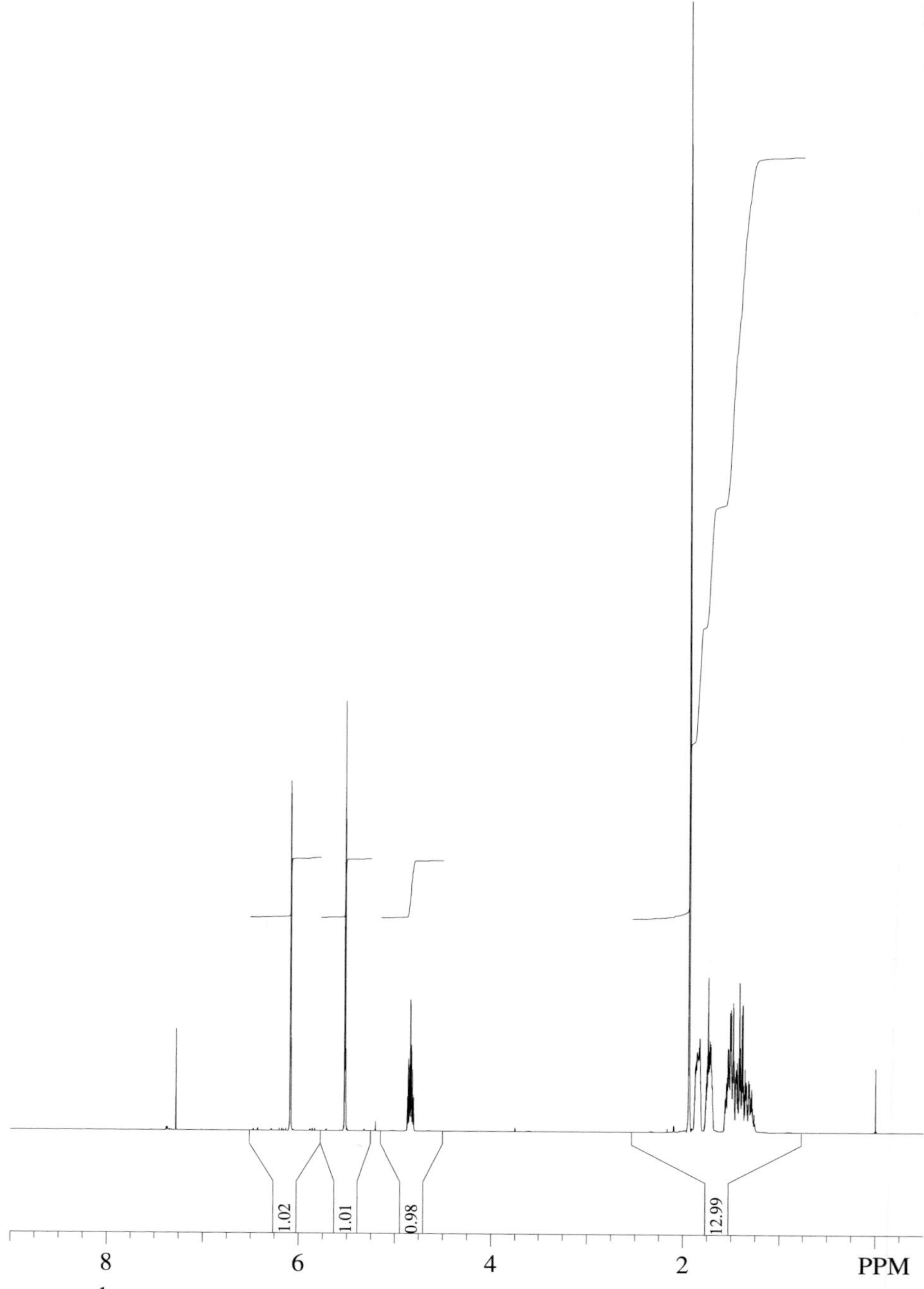

Fig. 6.3: ^{1}H-NMR spectrum: 400 MHz, solvent: $CDCl_3$

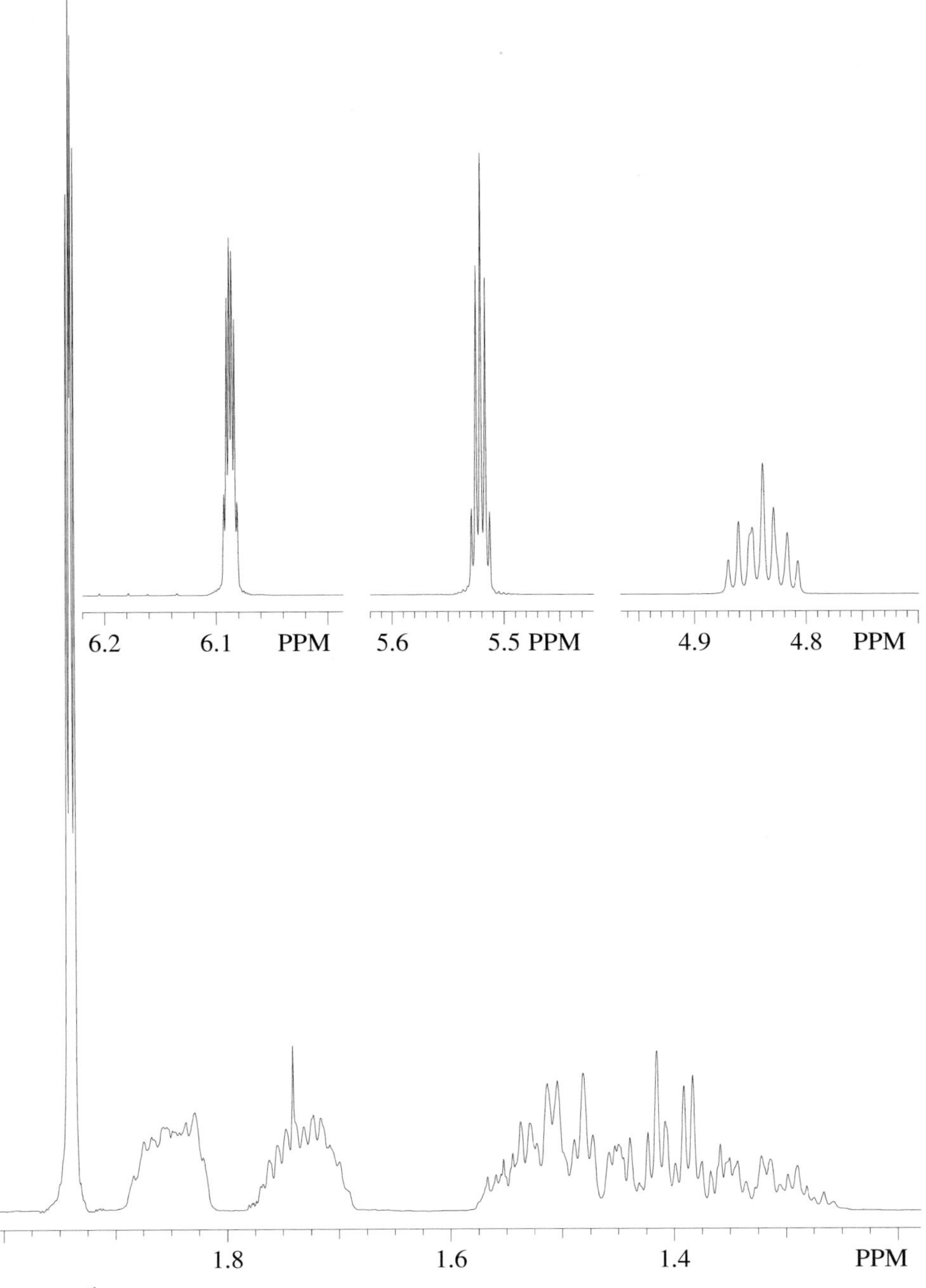

Fig. 6.4: ^{1}H-NMR spectrum (extended regions): 400 MHz, solvent: $CDCl_3$

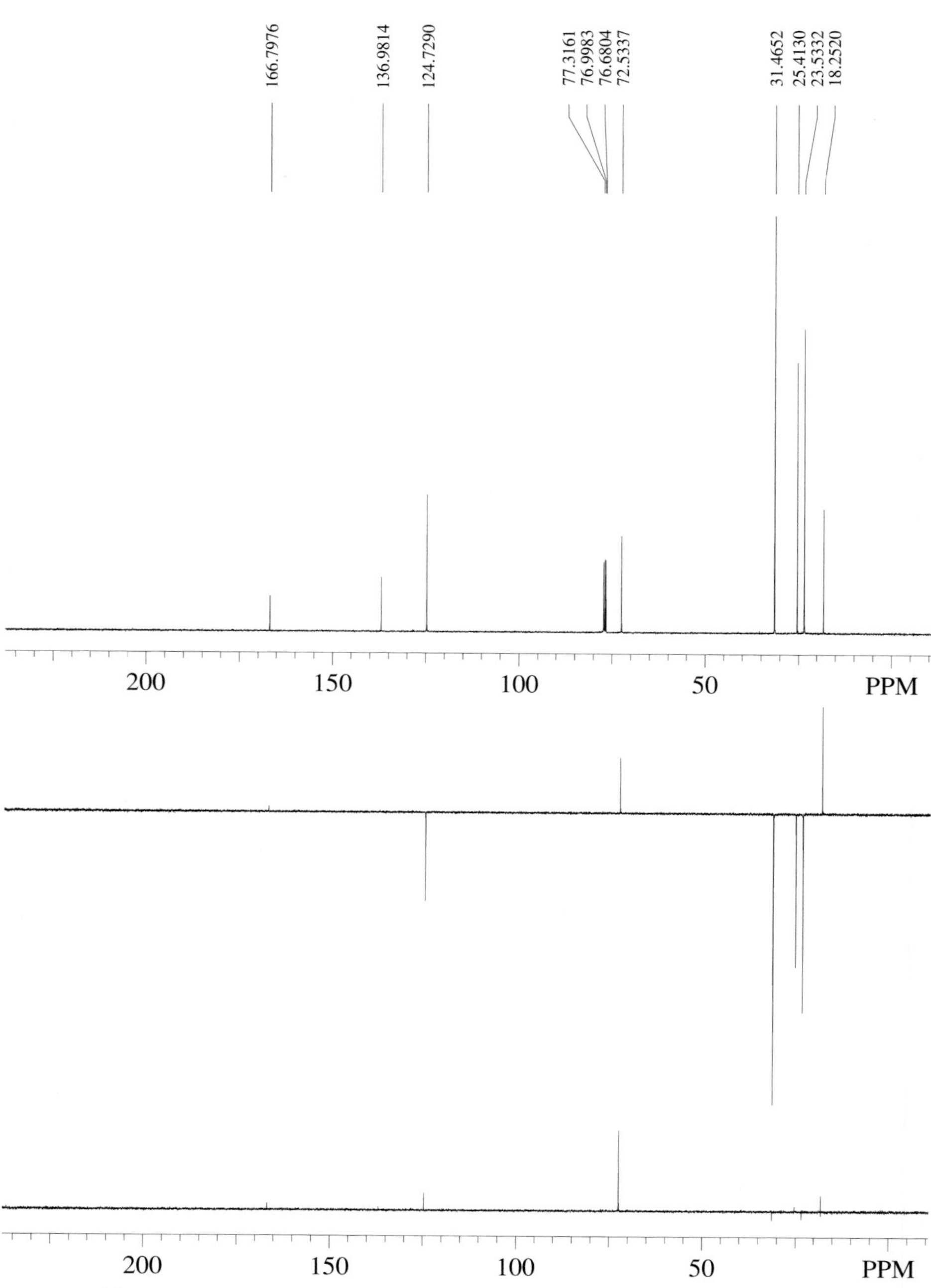

Fig. 6.5: ^{13}C-NMR spectrum: 100 MHz, solvent $CDCl_3$, Top: proton decoupled, middle: DEPT135, bottom DEPT90 ($\tau = 3.6$ ms)

6.1 Elemental Composition and Structural Features

The infrared spectrum exhibits a strong absorption band in the carbonyl region at 1710 cm^{-1} and a band at 1640 cm^{-1}, which is assigned to a carbon-carbon double bond. Due to the relatively high intensity of this band we expect the double bond being either linked to a hetero atom, or *cis* disubstituted, or having a terminal methylene group.

In the proton NMR spectrum, the three signals at $\delta = 6.1$, $\delta = 5.5$, and $\delta = 4.8$ correspond to one proton each. The remaining signals are difficult to separate. Together they correspond to 13 protons resulting a total of 16 hydrogen atoms in the molecule. As judged from their chemical shift values, the first two signals are due to two olefinic protons. As they only show very small splittings, they are most likely to be bonded to the same carbon atom in a terminal vinylidene group, in line with infrared spectral evidence. A quick look at the carbon-13 NMR spectra confirms this assumption as we find one signal for CH_2 and one for C in the appropriate shift range at $\delta = 124.7$ and $\delta = 137.0$, respectively. In the proton NMR spectrum, the signal at $\delta = 4.8$ shows a fairly complicated coupling pattern. We assume a methine group bonded to at least two proton bearing substituents (to explain the coupling) and to an oxygen atom (to explain the chemical shift value).

The carbon-13 NMR spectrum confirms the presence of a carbonyl group by exhibiting a signal at $\delta = 166.8$. From infrared evidence, a ketone would fit the observed frequency very nicely. However, the observed chemical shift positively excludes a ketone as we then would find the carbonyl signal near or above 200 ppm. This conflict is resolved by assuming an α,β-unsaturated ester.

The next two signals have already been assigned to the carbon atoms forming the carbon-carbon double bond. The line at $\delta = 72.5$ corresponds to the postulated methine group, its chemical shift value corroborating the assumption of a neighboring oxygen atom. The remaining four lines correspond to three methylene groups and one methyl group. The respective signals in the proton NMR spectrum are all between $\delta = 1$ and $\delta = 2$. As the proton NMR spectrum indicates 13 protons in this range, whereas we find evidence for only 9 protons in the corresponding part of the carbon-13 NMR spectrum, we must assume that some lines in the carbon-13 NMR spectrum coincide.

If we now add up all the structural fragments found so far, we get $C_{10}H_{16}O_2$ with a molecular mass of 168 which is at variance with mass spectroscopic evidence. The mass spectrum ends with a low intensity peak at m/z 169. The deviation is most easily rationalized by assuming protonation of the molecular ion which is not uncommon with esters. The spectrum is of the aliphatic-unsaturated type (m/z 27, 41, 55, 69), with one prominent even mass fragment ion at m/z 82. The fragments at m/z 150 and m/z 135 correspond to the consecutive loss of 18 mass units (water) and 15 (methyl radical) from the (unprotonated) molecular ion.

Check it in SpecTool:
Select **MS Tools, Homologous Mass Series** and enter one of the mass numbers from the ion series m/z 27, 41, 55, 69 ... and then click at the button: Find mass.

This brings up a list of the compound classes for which this ion series is characteristic.

6.2 Structural Assembly

We have so far identified the following fragments:

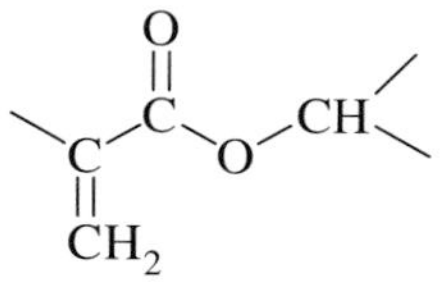

1 —CH_3

2 —CH_2—

2 —CH_2—

1 —CH_2—

We obviously have only one terminal function, namely the methyl group. This indicates that a ring must be present. If we calculate the double bond equivalents for the elemental composition found, we indeed obtain three. The methyl group cannot be placed at the methine group, as we then would have a doublet in the proton NMR spectrum. Thus, we place it at the double bond. We then expect a singlet around $\delta = 2$ in the proton NMR spectrum, broadened or split by long range coupling with the alkene protons. These protons in turn are split up with the same tiny coupling constant. The actual spectrum shows a doublet of doublets for the methyl group which is in line with our hypothesis.

We have now to place the five methylene groups on the methine group such as to (a) form a ring and to (b) create two pairs of isochronous carbon atoms. There is only one possible solution, which therefore represents the constitution of the unknown:

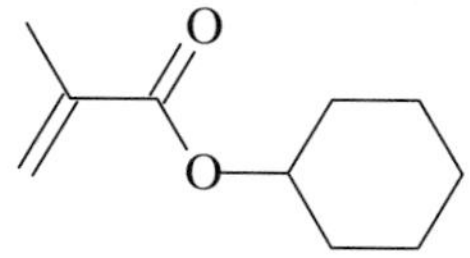

6.3 Comments

6.3.1 Mass Spectrum

The repeatability (same instrument, same operator, same day) of mass spectra is quite high. However the reproducibility (different instruments, different operators) can be surprisingly low. Intensities may change considerably in particular if the peak is produced by a bimolecular reaction, as e.g., protonation of the molecular ion. This is illustrated by the mass spectrum shwon in Fig. 6.6 of the sample which was recorded on a magnetic sector instrument rather than on a GC/MS combination using a quadrupole mass filter (Fig. 6.1).

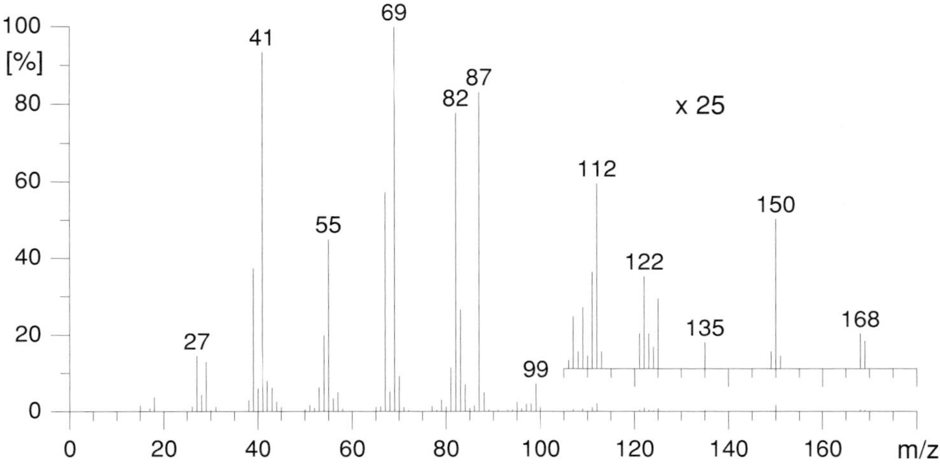

Fig. 6.6: Mass spectrum: EI, 70 eV

The mass spectra are dominated by two fragmentations common to esters, namely a double hydrogen rearrangement to form the protonated acid of m/z 87, which then loses water and carbonyl in succession (m/z 87 $\rightarrow$ 69 $\rightarrow$ 41), and an acid elimination from the molecular ion to yield the cyclohexene radical cation m/z 82 which subsequently loses (typically) methyl to a fragment C_5H_7 m/z 67. The remaining major features of the spectrum are accounted for by formation of cyclohexyl ion m/z 83 and by loss of ethylene from both m/z 83 and 82 to yield m/z 55 and 54, respectively.

6.3.2 Infrared Spectrum

The intensity of C=C stretching absorptions varies widely. In symmetrically *trans-* or in symmetrically tetrasubstituted double bonds the C=C stretching frequency becomes inactive in the infrared spectrum (it is, however, strong in the Raman spectrum). Even when the two or four substituents are not exactly alike, and in trisubstituted double bonds, the intensity can be quite low. Monosubstituted, *geminal* and *cis* disubstituted double bonds generally exhibit a medium intensity band. Highest absorptions are observed with double bonds directly bonded to an oxygen or nitrogen atom or in conjugation with a carbonyl group.

6.3.3 Proton NMR Spectrum

In monosubstituted hydrocarbons the chemical shifts of the protons next to the substituent increase in the sequence methyl, methylene and methine. For some non-spherical substituents e.g., –NHCOR and –OCOR the difference between the chemical shift values of methylene and analogous methine groups is especially large and extends to almost 1 ppm.

Check it in SpecTool:
Verify this with the data on page **Monosubstituted Alkanes** (button: CH_3-X, CH_3CH_2-X, etc.) in the **HNMR Data**, **Alkanes** section.

For the predominantly occurring conformation of the cyclohexyl ring, the methine proton is in axial position. Thus, it exhibits two diaxial (large) and two diequatorial (small) couplings and its signal should have a triplet of triplets structure in a first order spectrum.

Check it in SpecTool:
Vicinal Coupling Constants can be found under **Special** in the **HNMR Data** section.

Since the coupling partners are part of a higher order spectrum, splitting pattern and line widths are influenced by higher order effects. However, here we find the expected (partly overlapping) triplet of triplets.

The arrangement of the substituents in disubstituted ethylenes with two different substituents can easily be determined on the basis of the coupling constants. In ethylene the following values were measured: J_{gem} = 2.5 Hz, J_{cis} = 11.6 Hz, and J_{trans} = 19.1 Hz. With increasing electronegativity of the substituent(s) all three coupling constants decrease. The geminal coupling constant changes thereby its sign and its absolute value increases with further increase of the electronegativity of the substituent(s). In fluoroethylene the following values of proton-proton coupling constants were measured: J_{gem} = -3.2 Hz, J_{cis} = 4.7 Hz, J_{trans} = 12.8 Hz. The relative signs of the coupling constants do not influence the appearance of first order spectra but can be determined with the aid of double resonance techniques.

Check it in SpecTool:
Data for unsubstituted ethylene can be found under **HNMR Data**, **Alkenes**, **Unsubstituted Hydrocarbons**. The next page, accessible by clicking at the hand symbol pointing to the right, lists data for monosubstituted ethylenes. Click on the relevant table headings for coupling constants to display the entries in ascending numerical order and note that the more electronegative substituents cluster in the upper part of the list.

The absolute magnitudes of allylic coupling constants, which vary between 0 and 3.5 Hz, are mainly influenced by stereochemistry. *Cisoid* and *transoid* allylic coupling constants are generally of different magnitude. If the double bond is not an exocyclic one, generally *cisoid* coupling constants are somewhat larger than *transoid* coupling constants. In the present spectrum, 6 lines are observed for the low-field signal and 5 for the higher-field one. This can be understood by realizing that the geminal coupling constant is equal to twice the allylic one in the former case whereas geminal and allylic coupling constants are approximately equal in the latter one.

Check it in SpecTool:
To get detailed view on the splitting patterns use the expansion facilities in **SpecLib**. Line spacings can be easily read out by selecting one line, pressing the **Shift Key** and dragging the mouse.

The dependence of allylic coupling constants on the configuration and conformation is summarized under **HNMR Data**, **Alkenes**, **Allylic Couplings**. The signal at 6.1 ppm exhibits a smaller allylic coupling constant than the other one and can be assigned, therefore, to the proton in *trans* position relative to the methyl group.
Also the estimation of the chemical shifts allows for an unequivocal assignment: Access the **HNMR Tools**, **Shift estimation**: **Various Skeletons, Alkene**. Be sure to use COOR (isolated) as a substituent; COOR (conjugated) is used if the substituent has additional conjugations.

6.3.4 Ultraviolet Spectrum

The ultraviolet spectrum is well in line with the postulated structure.

Check it in SpecTool:
Estimate the UV absorption in **UV Tools**, **α,β-Unsaturated Carbonyl**, **Noncyclic parent system**. Place substituents and select the solvent. The estimated value of 203 nm agrees reasonably well with the measured one.

7 Problem 7

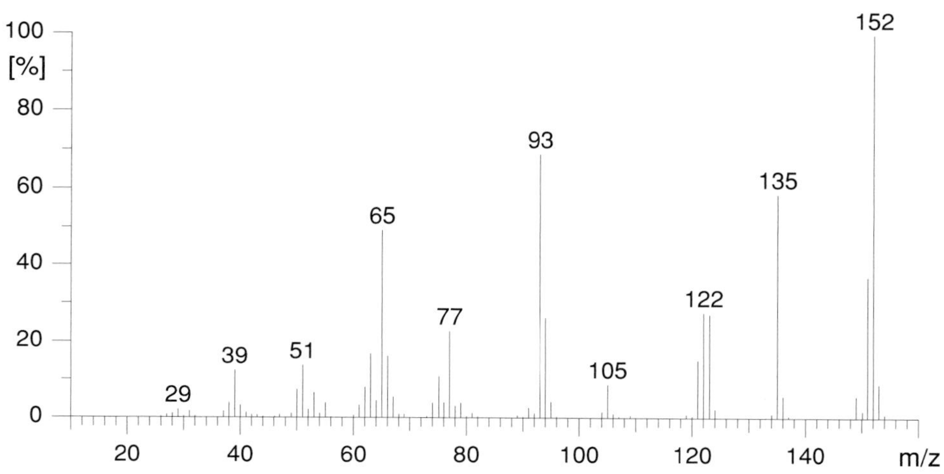

Fig. 7.1: Mass spectrum: EI, 70 eV

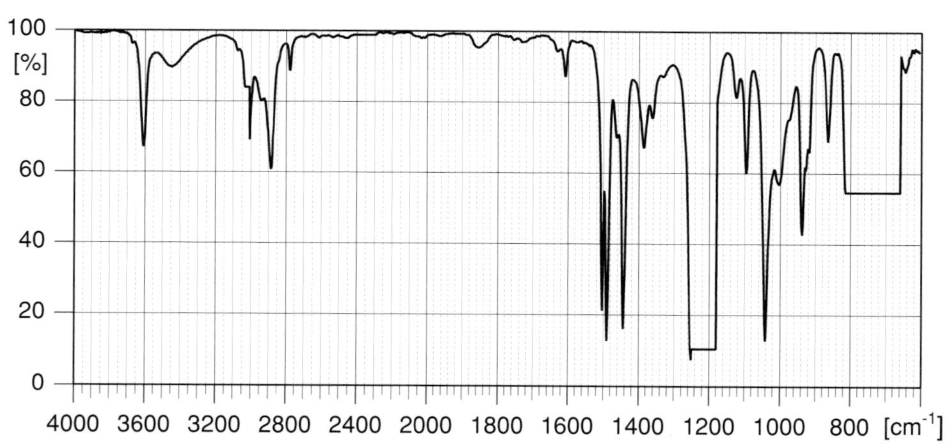

Fig. 7.2: IR spectrum: recorded in $CHCl_3$, cell thickness 0.2 mm

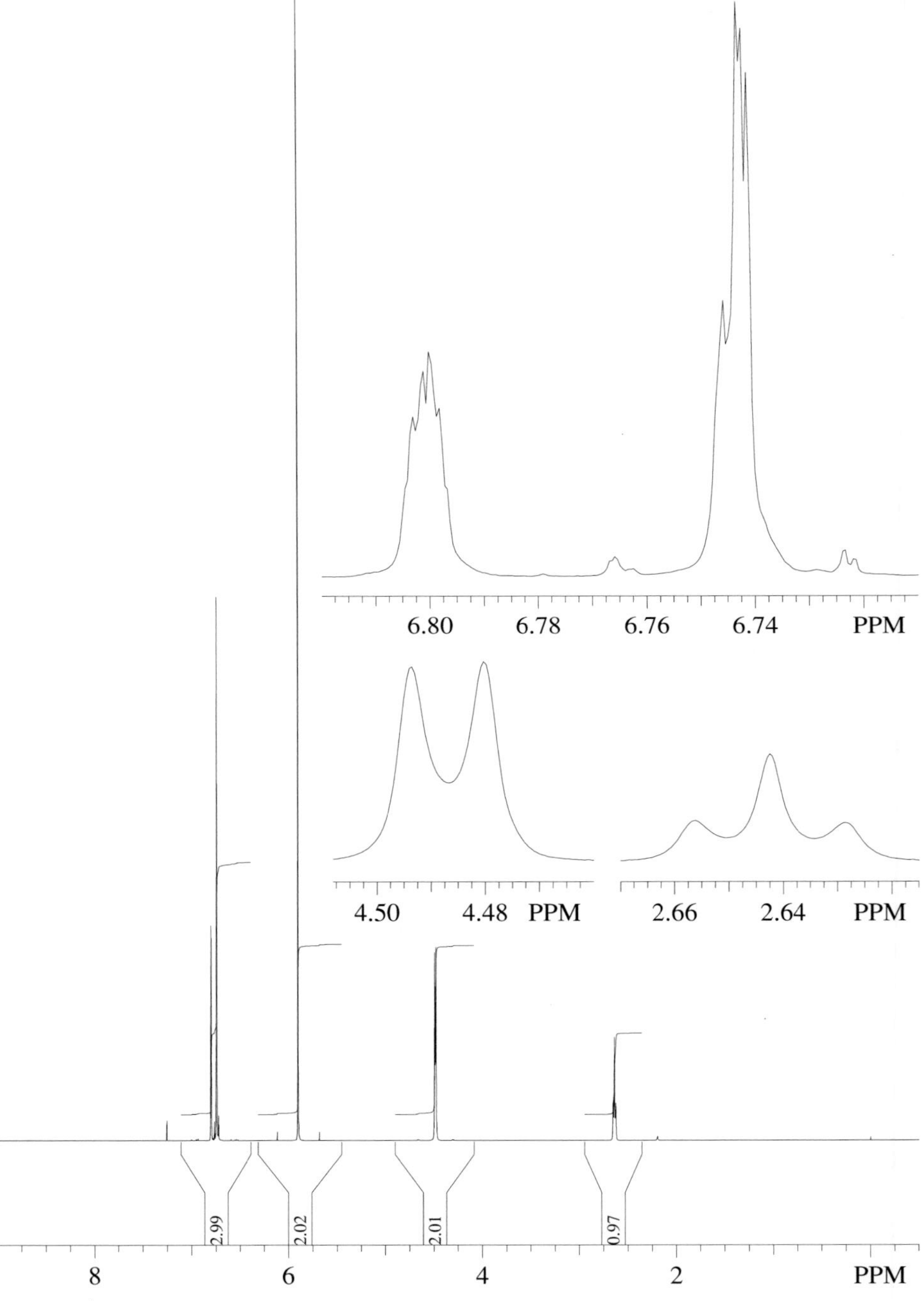

Fig. 7.3: ^{1}H-NMR spectrum: 400 MHz, solvent: $CDCl_3$

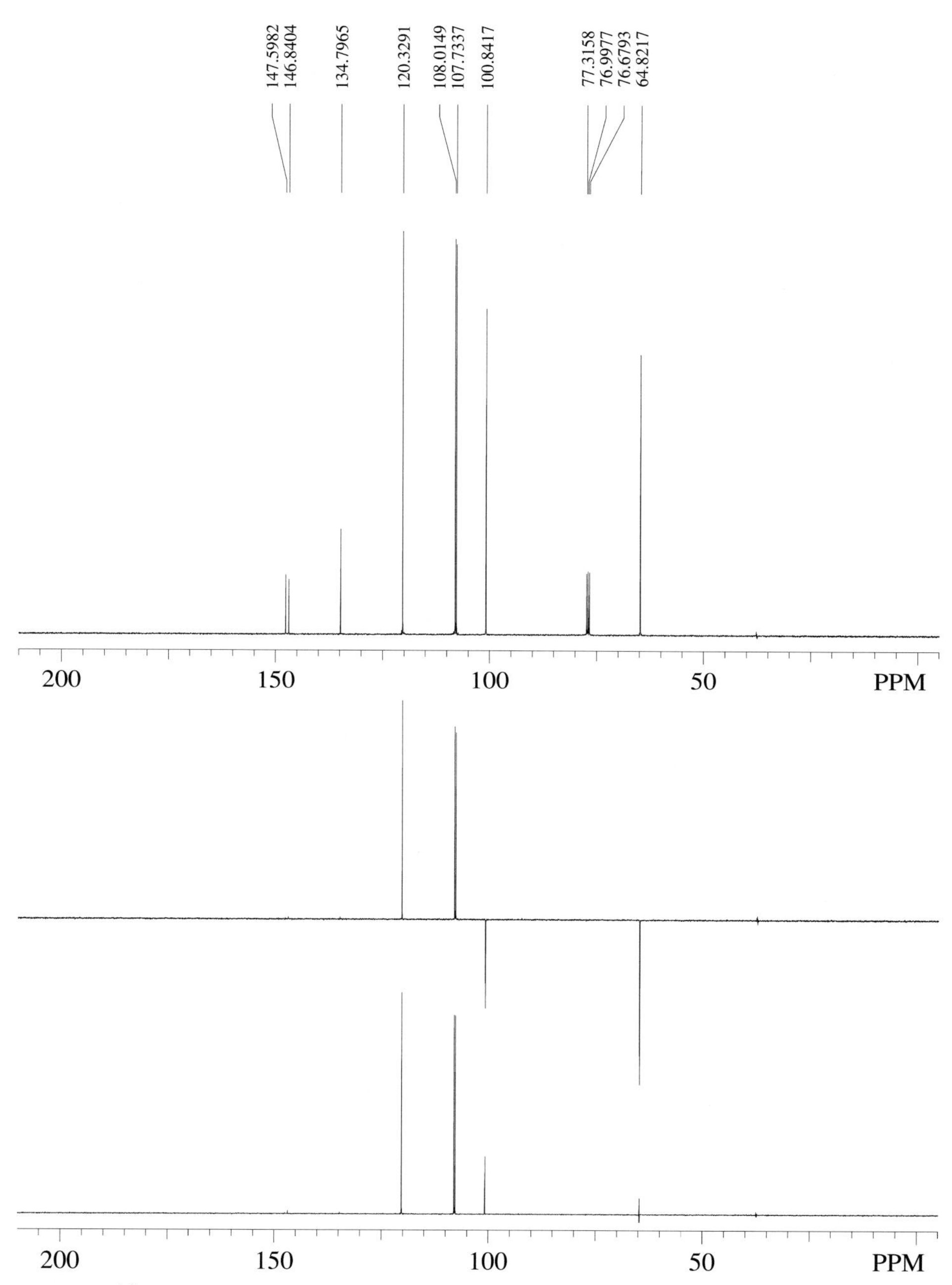

Fig. 7.4: ^{13}C-NMR spectrum: 100 MHz, solvent $CDCl_3$, Top: proton decoupled, middle: DEPT135, bottom DEPT90 ($\tau = 3.6$ ms)

7.1 Elemental Composition and Structural Features

The infrared spectrum shows clearly the bands characteristic for a hydroxyl group at 3600 cm^{-1} (OH stretching vibration free) and at 3450 cm^{-1} (OH stretching vibration associated). NH as possible assignment can be excluded, as NH_2 groups give rise to two bands of equal linewidths and NH groups never absorb at frequencies above 3500 cm^{-1}.

Check it in SpecTool:
Access **Primary** and **Secondary** amines in the **IR Data** section and inspect the prototype spectra.

We further note in the CH stretching region a band at 2770 cm^{-1}. We expect CH absorption bands at the lower end of the standard range for methyl groups and sometimes methylene groups bonded to nitrogen in amines or to oxygen in ethers, and for aldehydes. Furthermore, in alicyclic systems we often find weak combination bands in this region. The carbonyl region is empty. An aromatic moiety is indicated by the bands at 1600 cm^{-1} and 1500 cm^{-1}.

The mass spectrum is dominated by the peak at m/z 152, most probably corresponding to the molecular ion. The aromatic ion series (m/z 39, 51, 65, 77) accounts for all main peaks in the lower mass region. However, instead of the expected peak at m/z 91 or m/z 92 we find an intense peak at m/z 93. This could be indicative for a heteroatom in or on the aromatic system.

Check it in SpecTool:
Access **MS Data**, **Aromatic Compounds**, **Fragmentation Scheme**: The benzyl-tropylium tautomerism makes atoms on the ring and atoms in the ring equivalent.

The integration in the proton NMR spectrum gives a proton ratio of (from left to right) 3 : 2 : 2 : 1. The carbon-13 NMR spectrum shows eight signals. The DEPT spectra indicate that the lines at $\delta = 64.8$ and at $\delta = 100.8$ correspond to CH_2 groups and those at $\delta = 120.3$, 108.0, and 107.7 to CH. This corresponds well to the proton NMR spectrum in which the signal at 2.64 ppm can now be assigned to the hydroxyl proton identified in the IR spectrum. The line at $\delta = 64.8$ corresponds with the signal at $\delta = 4.5$ in the proton NMR spectrum and is assigned to a methylene group. Both the chemical shifts and the coupling of the protons to that at 2.64 ppm show that the methylene group is attached to the hydroxyl group.

A CH_2 group at $\delta = 100.8$ normally calls for a terminal methylidene group. This hypothesis could be in line with the corresponding signal in the proton NMR spectrum at $\delta = 5.9$. However, there is no evidence whatsoever for a carbon-carbon double bond in the infrared spectrum. Furthermore, the fact that the signal at $\delta = 5.9$ is a sharp singlet requires equal chemical shift values for both protons, a rather unlikely coincidence for methylidene protons in a molecule not possessing a high degree of symmetry. A rough comparison of the approximate molecular mass due to the atoms already identified with the actual molecular mass positively excludes a highly symmetric constitution. A chemical shift of about $\delta = 100$ for a methylene group may, however, also be induced by substituting it with two oxygen atoms. The resulting methylenedioxy group would also explain the chemical shift in the proton NMR spectrum and is further corroborated by the weak, albeit distinct infrared band at 2770 cm^{-1}. There are now six unassigned signals left in the carbon-13 NMR spectrum, three of them being CH groups. This agrees well with the three protons at $\delta = 6.7$ - 6.8 left unassigned in the proton NMR spectrum. We thus postulate a threefold substituted benzene ring.

Summing up the fragments found so far (C_6H_3, O–CH_2–O, CH_2–OH), we arrive at a mass of 152 Daltons. Thus, we have found all fragments, and we are now left with the problem of connecting them.

7.2 Structural Assembly

There are three possibilities for placing any three substituents on a benzene ring, namely in the 1,2,3-, 1,2,4-, or 1,3,5-positions. The latter may be excluded here, as the methylenedioxy group must take two adjacent positions on the benzene ring. Thus only two possibilities remain:

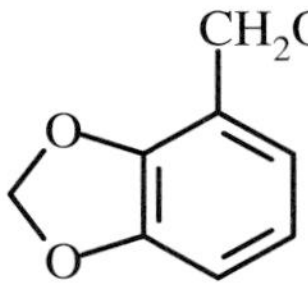

CH_2OH O O CH_2OH O O

I 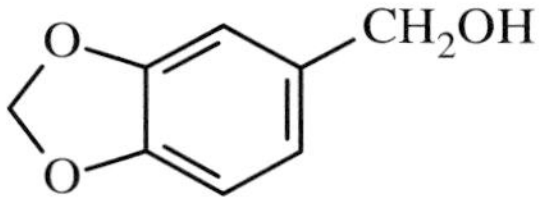 **II**

The decision between **I** and **II** is not an easy one. Infrared and mass spectra give no reliable information. The proton NMR chemical shifts predicted for **I** and **II** are not sufficiently different to allow for a decision. However, estimation of the carbon-13 NMR chemical shifts with additivity rules bring a solution forward.

Check it in SpecTool:
Access the additivity rule for substituted benzenes in the **CNMR Tools**, **Various Skeletons** section. An increment for the methylenedioxy group is not available. We thus substitute the increment for methoxyl. We may, therefore, expect quite some deviations in the chemical shift values. However, as we commit the same error for both tentative structures, the general pattern of the lines should not be greatly influenced. Estimate the shifts for both constitutions and compare the resulting patterns with the measured spectrum.

As expected, significant deviations between the predicted and measured values occur for both constitutions. However, they are significantly smaller for **II** than for **I** and, in addition, the general pattern showing two similarly deshielded CH carbon atoms for **I** and two similarly shielded ones for **II** only fits in the latter case the observed data. Thus we conclude that the unknown has the constitution **II**.

7.3 Comments

7.3.1 Mass Spectrum

Loss of formaldehyde (152 $\rightarrow$ 122, Δm = 30) followed by decarbonylation (122 $\rightarrow$ 94, Δm = 28) is characteristic of methylenedioxy groups on aromatic rings. Degradation of the hydroxymethyl group in benzyl alcohols typically follows the sequence of loss of a hydrogen radical and decarbonylation (m/z 152 $\rightarrow$ 151 $\rightarrow$ 123). These fragmentation paths have been confirmed by independent experiments.

7.3.2 Infrared Spectrum

In general, CH stretching vibrations around 3000 cm^{-1} are of low diagnostic value. There are, however, some partial structures that exhibit uncommon bands in this region. On the lower limit of the standard range, below 2850 cm^{-1}, we find absorption bands for methoxyl groups in ethers (but not in esters). Similar frequencies are found for methyl groups and often also methylene groups bonded to a nitrogen atom in amines (but not in amides). Furthermore, uncommonly low CH stretching frequencies are exhibited by methylenedioxy groups and by the dioxymethine group in acetals. In addition, absorption bands may be found in the same region for aldehydes and for cyclohexane moieties.

The use of the absence of a C=C stretching vibration band around 1650 cm^{-1} as an argument against the presence of a double bond is generally unreliable as the intensities can be very low. However, if the double bond is asymmetrically substituted so that the dipole moment changes during vibration, at least a medium intensity band is expected. Thus, exclusion of a terminal methylene group is valid.

7.3.3 Proton NMR Spectrum

The chemical shift and signal width of hydroxyl protons is highly dependent on solvent, temperature and impurities which contain exchangeable protons. Because of the different possible influences, the shift values are not characteristic of molecular environment in nonpolar solvents. The signal is often slightly broadened. The chemical shift of the hydroxyl protons in aliphatic alcohols without strong hydrogen bonds lies generally at higher field than δ = 5, mostly in the range of 1 to 2.5 ppm. They are, however, characteristic if dimethyl sulfoxide is applied as a solvent (see Problem 4).

The decision between **I** and **II** would be easy if the chemical shift values of the aromatic protons were sufficiently different to allow determination of their coupling constants. The coupling pattern would allow an unambiguous assignment since only **I** has a proton with two *ortho* couplings (about 8 Hz) and only **II** has a proton with *meta* coupling (ca. 2 Hz) without *ortho* coupling. *Para* couplings are usually not resolvable under standard experimental conditions. In the actual spectrum two protons at 6.75 ppm exhibit very similar chemical shifts so that the following analysis using first order rules must be confirmed by computer simulation. The small signals about 8 Hz apart from the central ones at 6.74 ppm (in a 400 MHz spectrum 0.02 ppm correspond to 8 Hz) indicate an *ortho* coupling between these protons. The total width of the signal at 6.80 ppm is only 3-4 Hz so that this proton cannot have any *ortho* coupling and, therefore, structure **I** can be excluded.

7.3.4 Carbon-13 NMR Spectrum

The DEPT90 spectrum should not show any signal for CH_2 groups. In the present case a fairly intense peak is observed at 100.8 ppm which seems to be in contradiction with the interpretation and also with the presence of a negative signal at this position in the DEPT135 spectrum. It must be kept in mind, that the simple interpretation of DEPT spectra is only valid if the timing fits the C–H coupling constant (i.e., $\tau = 1/2J$). The standard setting of $\tau = 3.6$ ms corresponds to a C–H coupling constant of 139 Hz. Due to the two oxygen substituents, J_{CH} is >160 Hz in the present case. This is the reason for the unexpected positive signal in the DEPT90 spectrum.

Check it in SpecTool:
Access the additivity rule for C–H coupling constants in the **CNMR Tools** section and estimate the value for a methylendioxy group. Compare it with the experimental one, which can be seen as a pair of small signals ("^{13}C-satellites"), symmetrically placed relative to the main peak at 5.9 ppm in the proton NMR spectrum. Use **SpecLib's** facilities for expanding the spectrum and precise evaluation of the magnitude of the splitting.

8 Problem 8

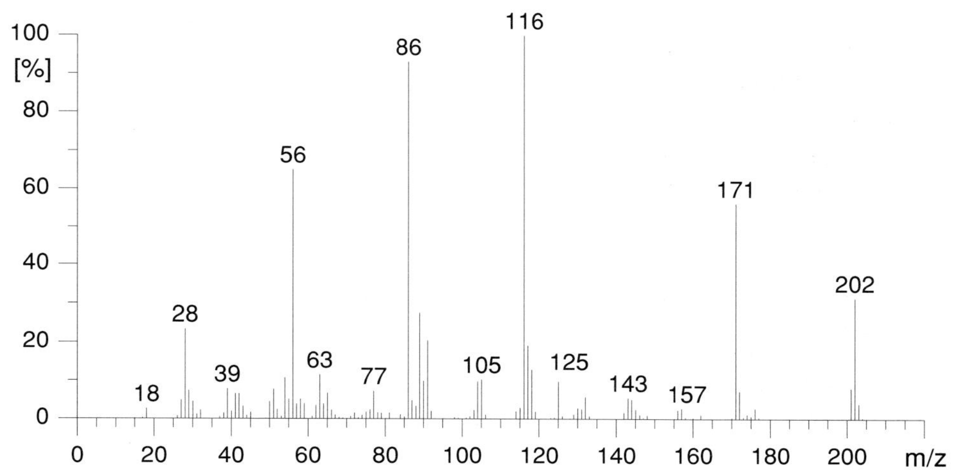

Fig. 8.1: Mass spectrum: EI, 70 eV

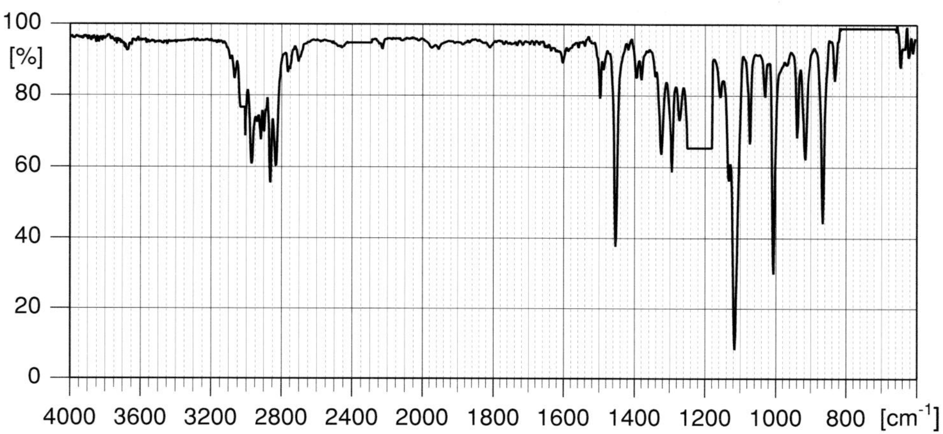

Fig. 8.2: IR spectrum: recorded in $CHCl_3$, cell thickness 0.2 mm

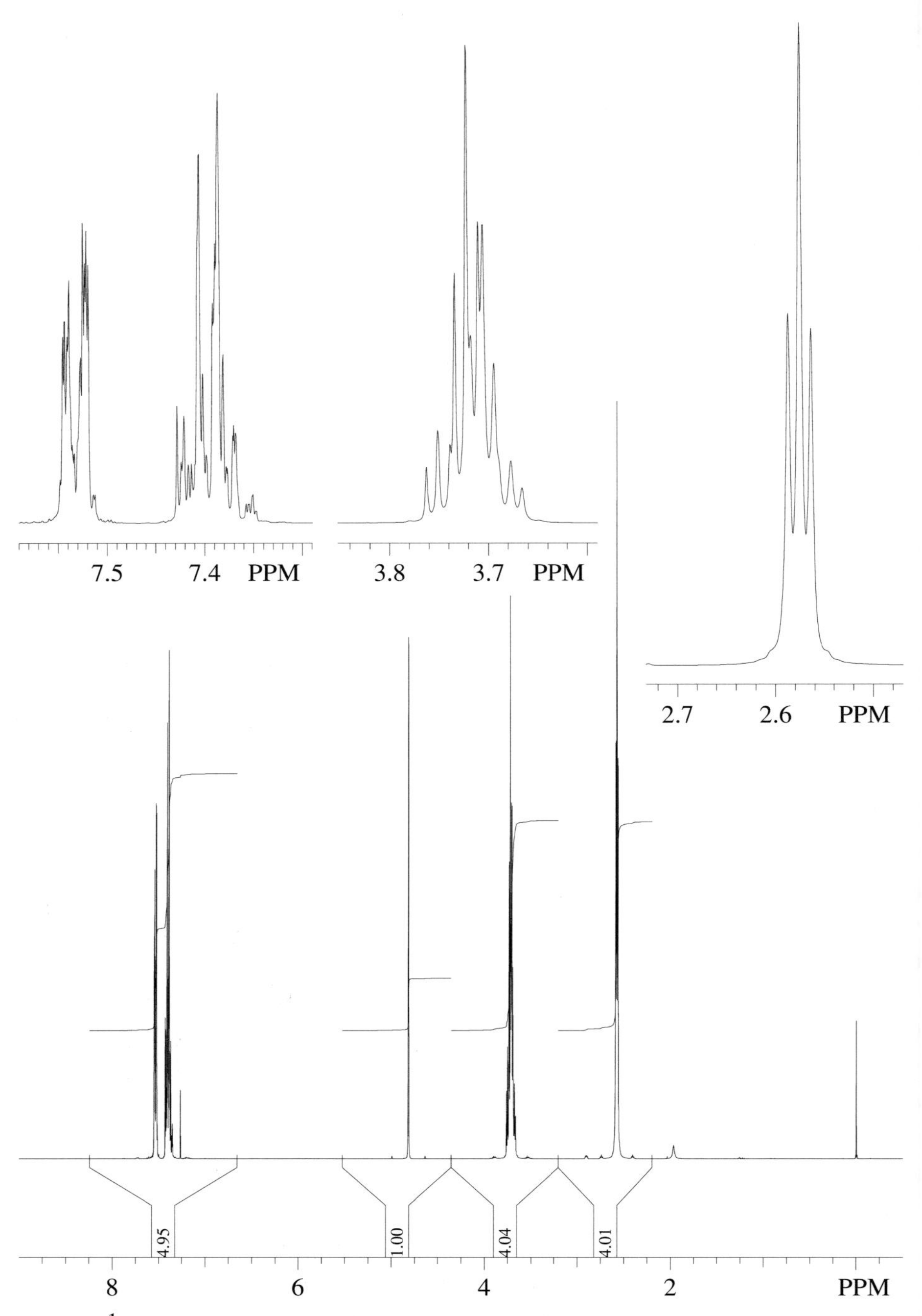

Fig. 8.3: ^{1}H-NMR spectrum: 400 MHz, solvent: $CDCl_3$

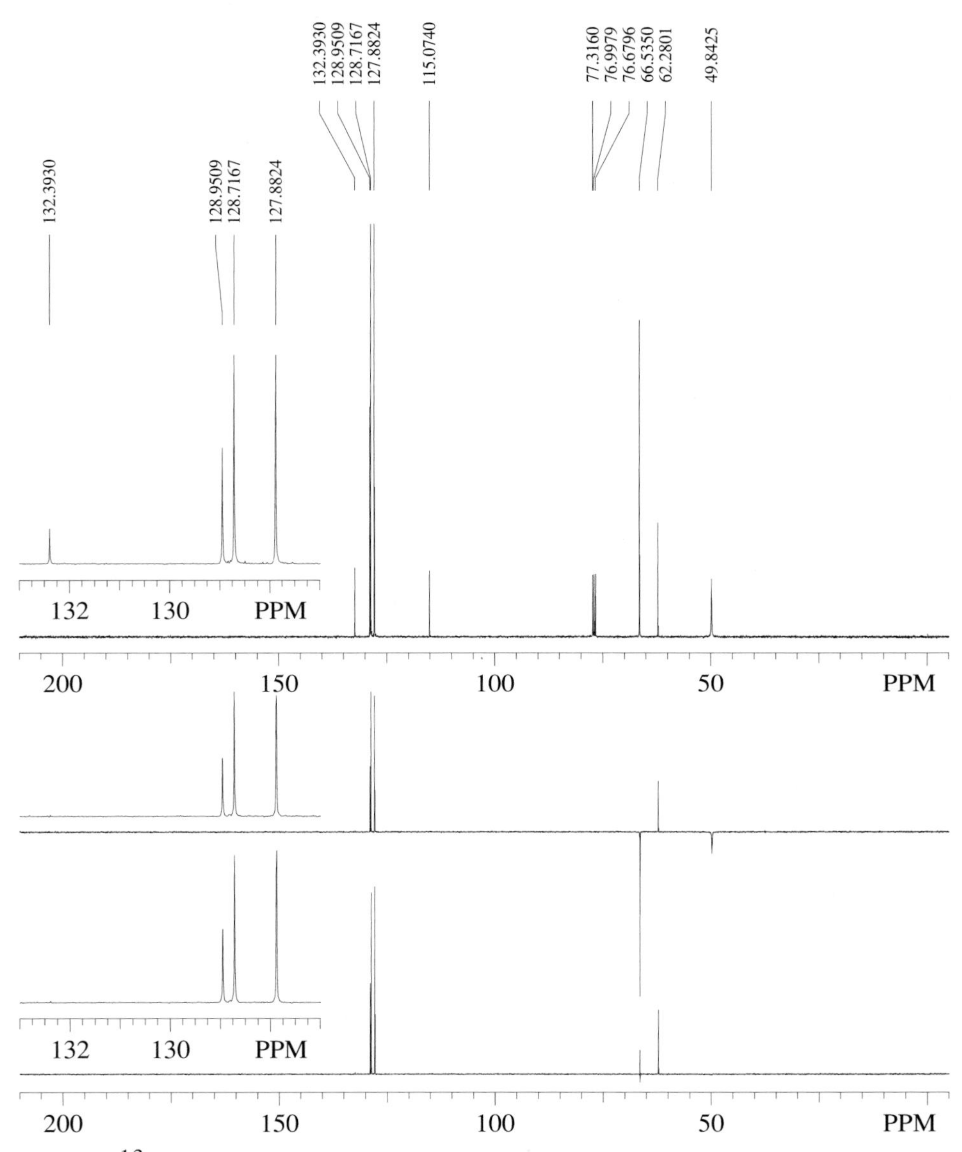

Fig. 8.4: ^{13}C-NMR spectrum: 100 MHz, solvent $CDCl_3$, Top: proton decoupled, middle: DEPT135, bottom DEPT90 ($\tau = 3.6$ ms)

8.1 Elemental Composition and Structural Features

Mass spectrum: m/z 202 is assumed to be the molecular ion, because all signals of lower mass can be related to it by chemically reasonable transitions. A carbon-13 isotope peak of 13% intensity relative to $M^{+\cdot}$ at m/z 203 limits the possible number of carbon atoms to twelve. The mass difference to the first fragment ion at m/z 171 indicates the presence of sp^3 hybridized oxygen atom. Fragment series m/z 39, 51, 63 to 65, 77, 91, 104, 105 and maxima at m/z 28, 56, 86 suggest a mixed aromatic/non-aromatic nature of the compound.

Infrared spectrum: C–H stretching vibrations above and below 3000 cm^{-1} and aromatic skeletal vibrations at 1500 cm^{-1} and 1600 cm^{-1} corroborate the conclusions drawn from the mass spectrum. No OH, NH or carbonyl groups are present.

Proton NMR spectrum: 14 protons with integral ratios of 2 : 3 (multiplet) : 1 (singlet) : 4 (multiplet) : 4 (multiplet) are evident. The five proton multiplet around δ = 7.4 – 7.6 may be assigned to a monosubstituted benzene ring, the singlet at δ = 4.8 to a methine group next to a hetero atom. In addition, there are two multiplets at 3.7 and 2.6 ppm corresponding to four protons each, which probably couple with each other.

Carbon-13 NMR spectrum: Four signals in the range of 100 to 132 ppm (three CH and one C) can be assigned to the monosubstituted benzene. There is one additional signal for a C at 115.1 or at 132.4 in this region. Additionally, we find a signal for a CH at 62.3 ppm and two for CH_2 groups at 66.5 and 49.8 ppm. According to the integration in the proton NMR spectrum, each of the latter ones corresponds to two methylene groups. The chemical shifts of the low-field ones (66.5 and ca. 3.7 ppm) require oxygen as one neighbor, and their fine structure shows that hydrogen atoms must be bonded to the second neighbor. The only available coupling partners are the other methylene groups (at 49.8 and ca. 2.6 ppm). Their chemical shifts shows that they must not be bonded to oxygen.

Check it in SpecTool:
Access the table of the monosubstituted alkanes in the **CNMR Data** section and click at an α-methylene group in the table heading. This will sort the table for increasing chemical shift values. Now click at 50 ppm in the distribution diagram below the table in order to scroll it to the relevant part and note that in this shift region no oxygen substituents can be found. Now click at 65 ppm to note that this chemical shift requires a direct connection of the methylene group to an oxygen atom. Switch to **HNMR** and confirm the assignments by the same procedure.

So far, we found the partial formula $C_{12}H_{14}O$ corresponding to a mass of 174 Daltons. The missing 28 mass units can either be accounted for by two nitrogen atoms or by an additional C and O leading to two possible molecular formulas, namely $C_{12}H_{14}N_2O$ and $C_{13}H_{14}O_2$, each with seven double bond equivalents. No convincing indications are easily available in favor of one of these formulas. However, no reasonable structure can be constructed for $C_{13}H_{14}O_2$ under the restrictions collected above (see Comments), which imply that two isochronous quarternary carbon atoms should be present. So the molecular formula is $C_{12}H_{14}N_2O$.

The signal at 115.1 ppm cannot belong to the monosubstituted benzene ring because only terminal substituents could induce such a chemical shift.

Check it in SpecTool:
To check the validity of the above statement, apply the same procedure as before to the table of monosubstituted benzenes in the CNMR section.

The chemical shift of $\delta = 115.1$ for the carbon signal requires it to be due to a cyano group, because no alternative assignment is possible under the given premises.

Check it in SpecTool:
Access **CNMR Ranges** directly from the **TOP** card and click at 115 ppm in the scale at the top of the page. By clicking into any of the then appearing bar diagrams you are taken in the respective data section.

Only an additional aromatic or heteroaromatic ring, a symmetrically persubstituted carbon-carbon double bond, or a carbon atom between two oxygen or halogen atoms could otherwise account for this chemical shift value.

8.2 Structural Assembly

According to the identified features the molecule consists of the following parts:

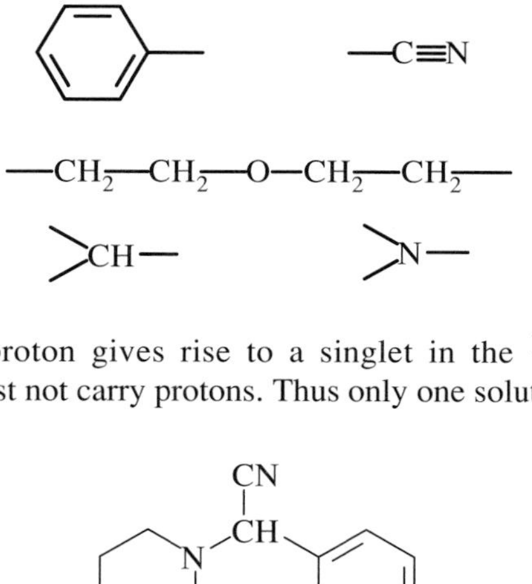

C_6H_5—   —C≡N

$—CH_2—CH_2—O—CH_2—CH_2—$

>CH—   >N—

Since the methine proton gives rise to a singlet in the ^{1}H-NMR spectrum, its immediate neighbors must not carry protons. Thus only one solution is possible, namely:

CN
|
N—CH—C_6H_5 (morpholine N, O)

8.3 Comments

8.3.1 General

An exhaustive generation of all possible isomers without using a corresponding computer program (structure generator) is notoriously unreliable and error prone. In the present case the structure generator ASSEMBLE [1] (not part of the enclosed version of SpecTool) was applied to generate all possible constitutions. Under the premises derived above only the following three isomers were generated for the excluded molecular formula $C_{13}H_{14}O_2$:

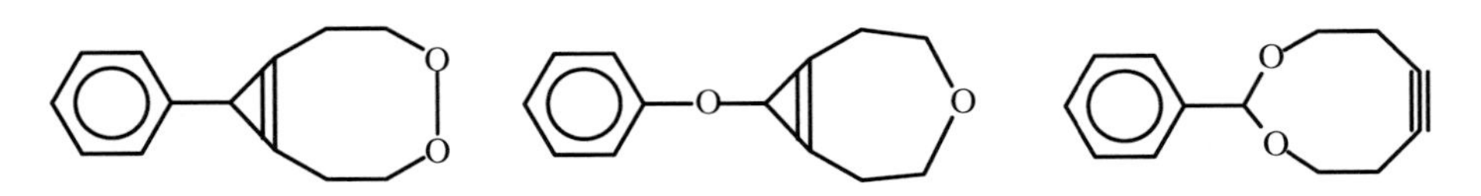

None of them is compatible with the experimental spectra.

Check it in SpecTool:
Estimate the ^{1}H- and ^{13}C-chemical shifts of the three structures generated by ASSEMBLE.

8.3.2 Mass Spectrum

In principle, the intensity of the carbon-13 isotope peak could be used to infer the number of carbon atoms in the molecule, as each carbon atom contributes 1.1% to the first isotope peak. However, the isotope peak may gain intensity from protonation of the molecular ion, thus simulating too many carbons. Therefore, only the an upper limit of the number of carbon atoms can be inferred. Furthermore, intensity values in mass spectra are of limited precision due to various instrumental and operating artifacts. Thus valid conclusions can only be drawn if the deviation between found and expected intensity of $[M+1]^+$ is large. In the present case, an intensity of 12.6% formally limits the maximum number of carbon atoms to 11. Both molecular formulas are within the tolerance.

The mass spectrum can be rationalized as consisting mainly of the two complementary halves of the molecule as obtained by benzilic cleavage (morpholine ring m/z 86, aromatic moiety m/z 116) and by loss of formaldehyde: (mass difference 30) from $[M-1]^+$ and from m/z 86 in a retro-Diels-Alder type reaction as follows:

+•
O N–R ⟶ $CH_2{=}CH$ $\overset{+}{N}$–R + $CH_2{=}O$
CH_2

R = —H
CN
—CH—(phenyl)

Loss of the phenyl group from the molecular ion yields m/z 125.

8.3.3 Infrared Spectrum

A nitrile group may be expected to give rise to a sharp signal in the range of 2230 to 2260 cm^{-1} and is in general easily identified. There is in fact a signal observable at 2230 cm^{-1}, but it is too weak in this case to be recognized as indicating nitrile at a first glance, especially because overtone and combination bands of similar intensity are present in its neighborhood and because it might as well be interpreted as an overtone of the strong band at 1120 cm^{-1}. It is a rather general observation, that nitrile absorptions are scarcely visible if the group is attached to a carbon atom carrying an electronegative substituent.

The various low intensity bands between 2500 cm^{-1} and 1500 cm^{-1} may be rationalized as follows. The broad band centered at 2450 cm^{-1} is an instrumental artifact, due to solvent absorption at this frequency. The band at 2230 cm^{-1} is assigned to the CN stretching vibration of the nitrile group. The double bands near 1970 cm^{-1} and 1810 cm^{-1} are due to various combination vibrations and overtones of the benzene moiety. The band at 1600 cm^{-1} finally is a skeletal vibration of the benzene ring. The very strong and comparatively sharp band at 1120 cm^{-1} is due to a vibration of the morpholine part of the molecule, which is probably best described as C–O–C asymmetric stretching vibration.

8.3.4 Proton NMR Spectrum

Since the molecule is chiral the geminal methylene protons are, in principle, anisochronous even if there is a fast ring inversion on the corresponding NMR time scale. A fast rotation around the CH–N bond together with a fast ring inversion cause an equivalence of each of the protons of one pair of methylene groups with the counterpart on the other pair of methylene groups (see Chapter 19.4). Thus, in the general case, if fast rotation and ring inversion occur, an *AA'BB'CC'DD'* spin system is to be expected. This is simplified to two identical *ABCD* spin systems if all couplings across the hetero atoms vanish. The actual spectrum clearly shows the non-equivalence for the low field protons, while the triplet of the high field methylene groups is a consequence of an accidental equivalence.

8.3.5 Carbon-13 NMR Spectrum

There are 8 signals in the carbon-13 NMR spectrum because two pairs of aromatic CH and of morpholino CH_2 are isochronous. Although the molecule is chiral, a fast rotation of the phenyl and morpholino groups causes chemical shift equivalence of the permutated atoms and one would naively expect double intensities. However, differences in relaxation times and nuclear Overhauser effect often bias line intensities (see Chapter 19.3). Nevertheless, identification of the *para* carbon atom of the benzene ring is obvious. For the morpholino group, the situation is more complicated because the axis of rotation becomes a symmetry element only due to fast conformation changes of the six-membered ring. One of the processes is slow enough to cause line-broadening of the signal at 49.8 ppm, resulting in reduced signal height. The integrated intensity, however, would probably be as expected, i.e., the same as for the other CH_2 signal.

By the same token, one cannot decide whether one of the signals of the quaternary carbons corresponds to one or two carbon atoms. Therefore, one cannot distinguish between the two empirical formulas $C_{12}H_{14}N_2$ and $C_{13}H_{14}O_2$ on the basis of the carbon-13 NMR spectrum.

8.4 References

[1] C.A. Shelley, M.E. Munk, Anal. Chim. Acta **133** (1981) 507.

9 Problem 9

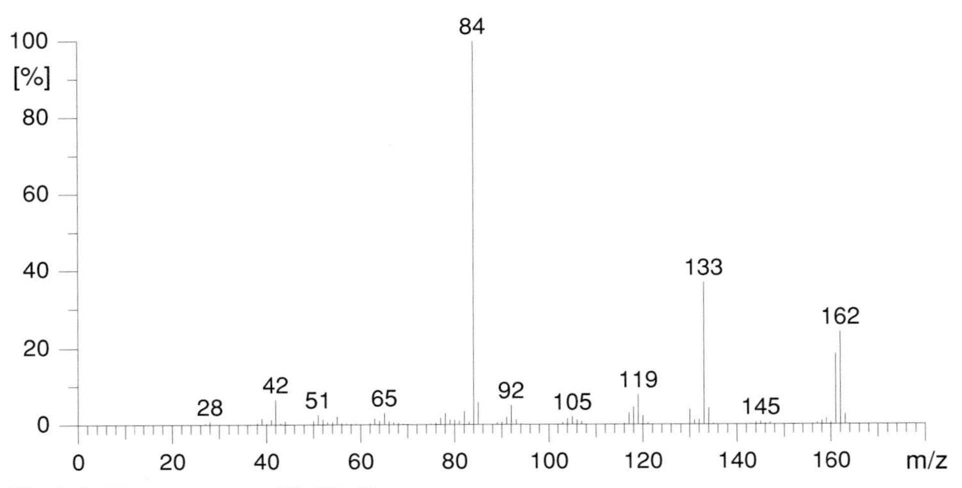

Fig. 9.1: Mass spectrum: EI, 70 eV

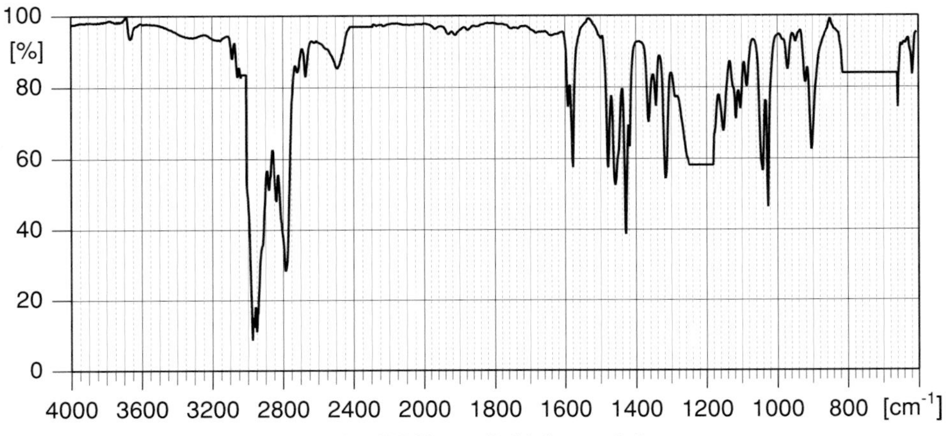

Fig. 9.2: IR spectrum: recorded in $CHCl_3$, cell thickness 0.2 mm

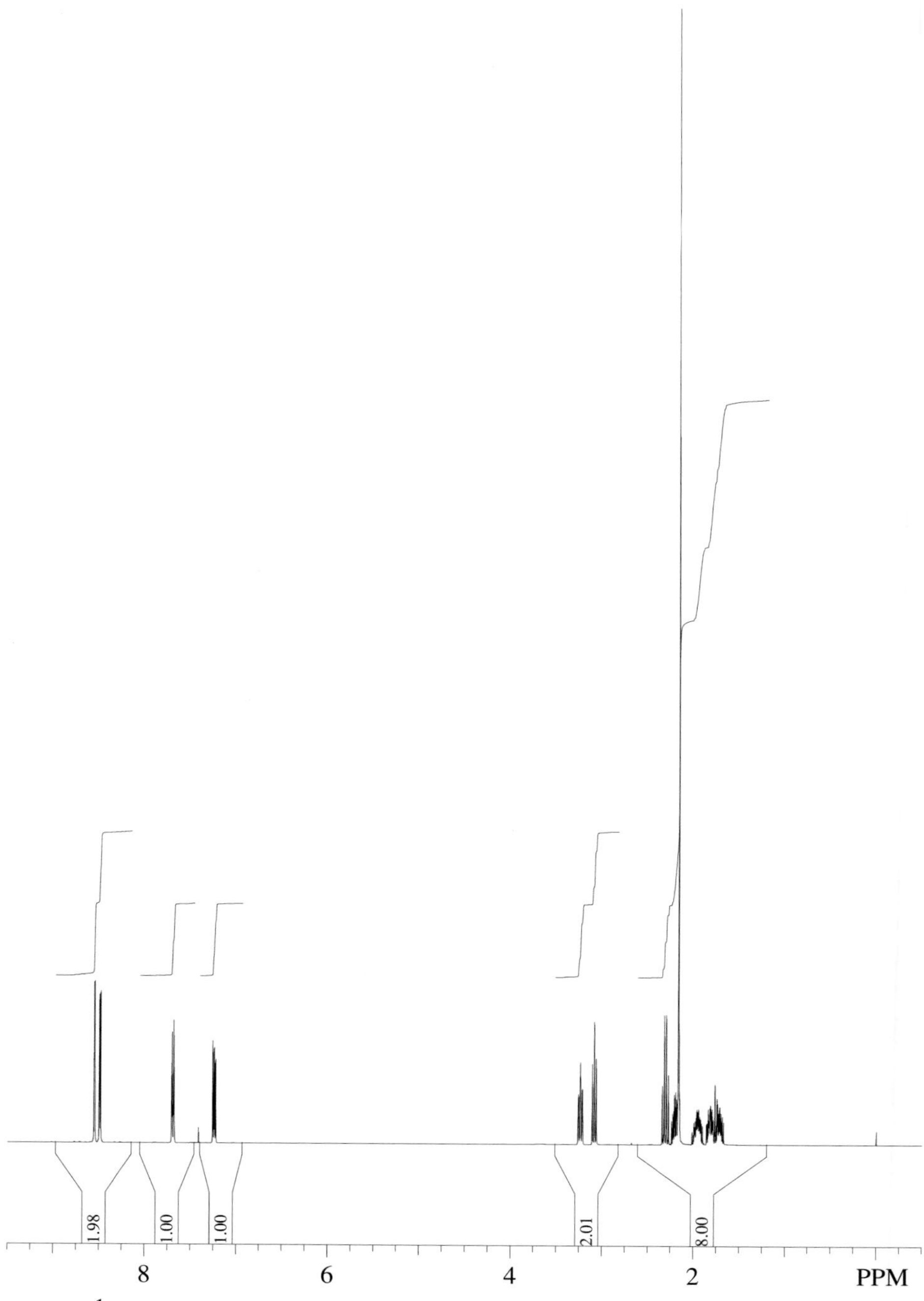

Fig. 9.3: ^{1}H-NMR spectrum: 400 MHz, solvent: $CDCl_3$

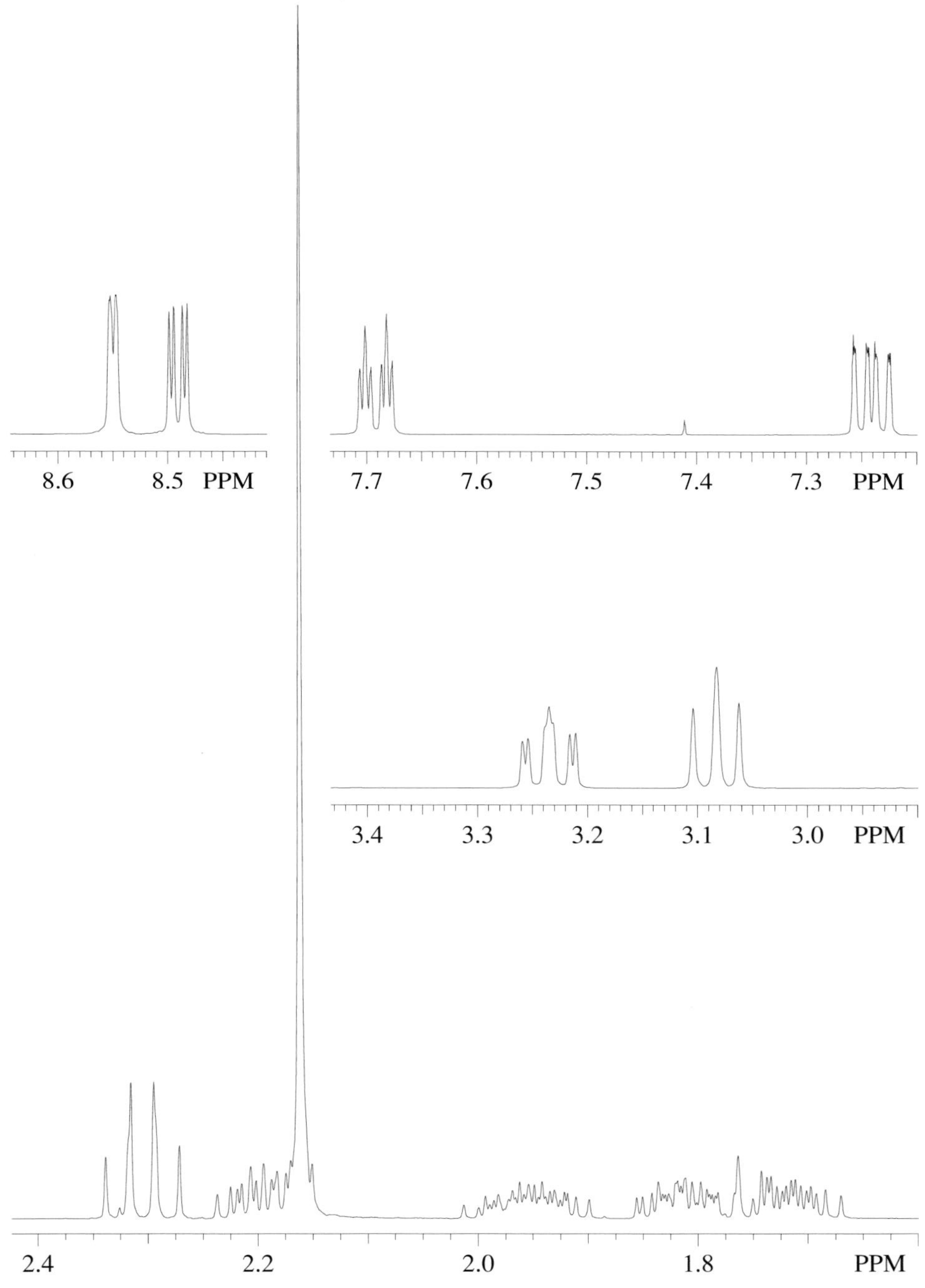

Fig. 9.4: ^{1}H-NMR spectrum (expanded regions): 400 MHz, solvent: $CDCl_3$

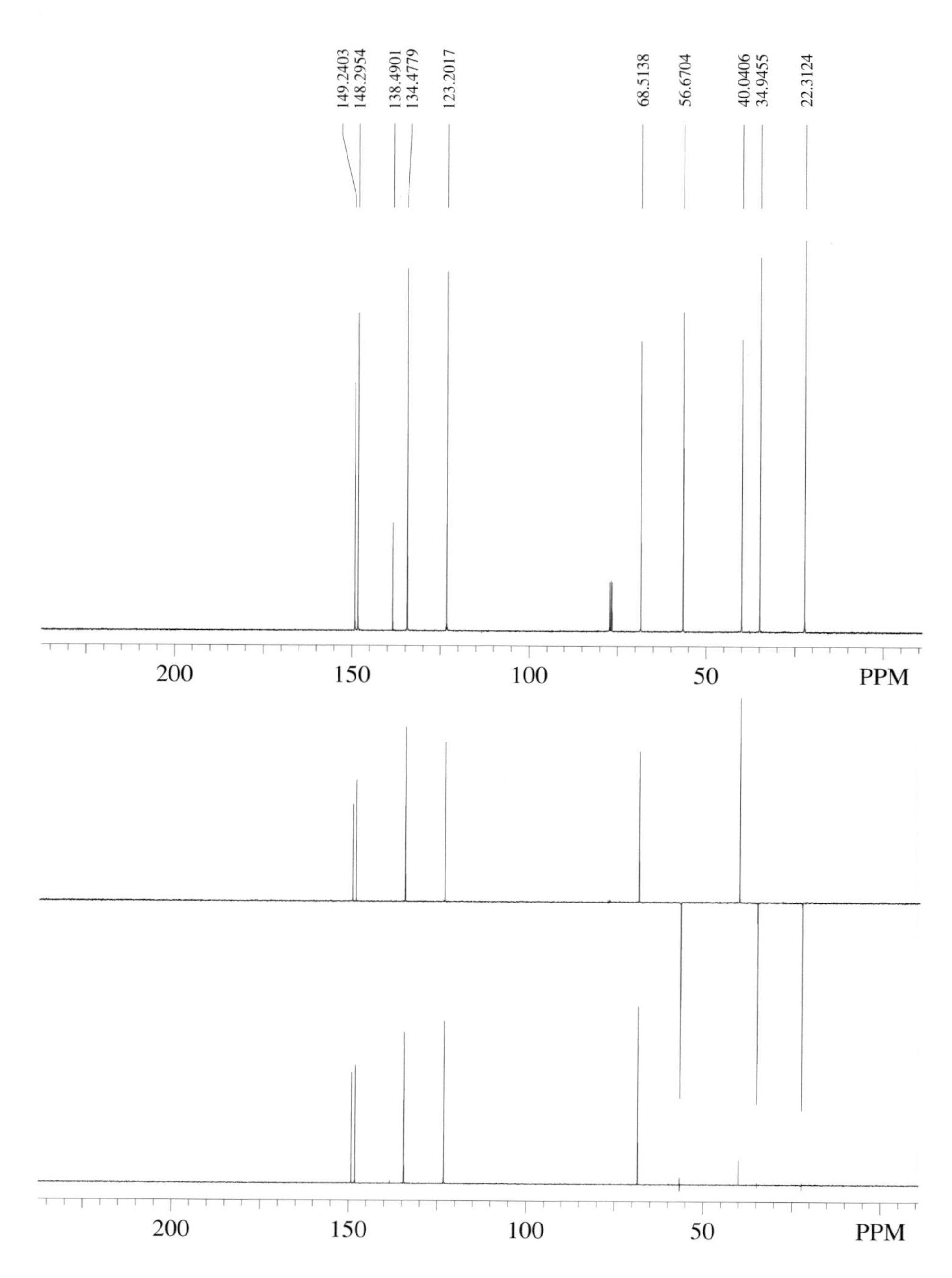

Fig. 9.5: ^{13}C-NMR spectrum: 100 MHz, solvent $CDCl_3$, Top: proton decoupled, middle: DEPT135, bottom DEPT90 ($\tau = 3.6$ ms)

9.1 Elemental Composition and Structural Features

In the infrared spectrum, we note two uncommon bands at 2680 and 2500 cm^{-1}. The signal at 1590 cm^{-1} could indicate an aromatic moiety.

> **Check it in SpecTool:**
> Inspect the prototype spectrum for aromatic hydrocarbons as given in **IR Data**, **Aromatic Compounds**, **Aromatic Hydrocarbons** and the corresponding **Spectra** section for specific examples (directly accessible via the **Switch Menu** or **Switch Palette**).

Note that most of the typical bands around 1220 cm^{-1} and 720 cm^{-1} are hidden by solvent absorption.

From the mass spectrum, we assume a molecular mass of 162. The most abundant peaks at m/z 133 and 84 correspond to a loss of 29 and 78 Daltons, respectively. The even mass numberof the base peak means that it is either produced with a rearrangement or that it contains an odd number of nitrogen atoms. In this latter case the molecule has to contain an even number of nitrogen atoms to explain the even molecular mass.

Ten signals are discernible in the carbon-13 NMR spectrum, indicating at least ten carbon atoms (if no signal splitting due to coupling with nuclei other than protons occurs). Five signals fall in the range expected for aromatic carbons (δ = ca. 100 to 160). Two of them exhibit rather extreme chemical shift values close to 150 ppm. We may thus tentatively assume that they are substituted with hetero atoms. Furthermore, both carbon atoms carry a hydrogen atom, as indicated by the DEPT spectra. Thus, if the assumption about the heteroatom substitution is correct, the hetero atom is part of the aromatic moiety. The DEPT spectra indicate five methine groups (four aromatic, one aliphatic), three methylene groups, one methyl group, and one quaternary (aromatic) carbon atom. We thus expect at least 14 hydrogen atoms in the proton NMR spectrum.

Integration of the signals in the proton NMR spectrum gives a hydrogen ratio of 1 : 1 : 1 : 1 in the aromatic region (δ = 6.5 to 9.0), and 1 : 1 : 1 : 4 : 1 : 2 in the aliphatic region, summing up to 14 hydrogen atoms. This is consistent with the information from the carbon-13 NMR spectrum. Within the group of aliphatic protons we easily identify a sharp singlet, corresponding to the methyl group already inferred from the carbon-13 NMR spectrum.

If we now sum up the elemental composition given by the partial structures identified so far, we arrive at $C_{10}H_{14}$ and at least one hetero atom. Thus, the difference to the molecular mass amounts to 162 - 134 = 28, which includes at least one hetero atom. There is not much choice as to the assignment of this difference: C_2H_4 is excluded because it lacks a hetero atom, for CO we would need an additional line in the carbon-13 NMR spectrum, so N_2 is the only remaining possibility. Thus, the elemental composition is $C_{10}H_{14}N_2$.

9.2 Structural Assembly

We now proceed to assemble a tentative constitution from our fragments. We expect a heteroaromatic system with five carbon atoms. In the present context this calls for a pyridine ring. As four aromatic protons are present, it must be monosubstituted. This is corroborated in the mass spectrum by the prominent loss of 78 (corresponding to C_5H_4N) from the molecular ion. Benzene can be ruled out as a possible assignment for 78 because the postulation of a phenyl group contradicts carbon-13 NMR evidence. The other fragment of mass 84 has an elemental composition of C_5H_8N and contains one double bond equivalent and one methyl group. The methyl group, giving a singlet in the proton NMR spectrum at $\delta = 2.15$ has to be bonded to the nitrogen atom. As no other terminal groups are available, the nitrogen atom will be part of the ring. Furthermore, chemical shifts of $\delta = 68.5$ and 56.7 in the carbon-13 NMR spectrum for methine and methylene groups clearly indicate the following partial structure:

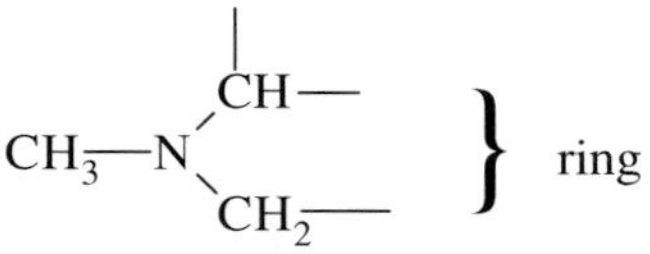

CH_3—N(—CH—)(—CH_2—) } ring

We may now construct all molecules not at variance with our inference:

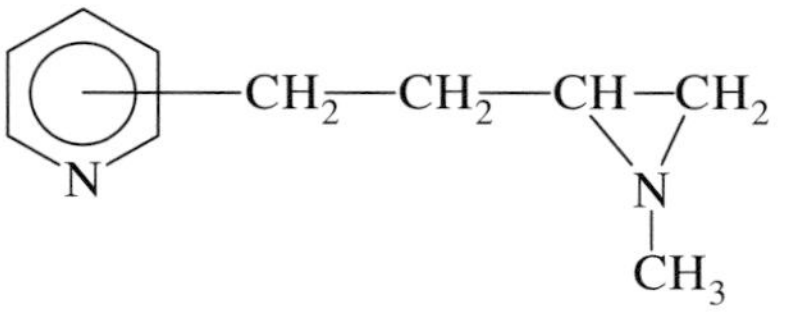

I

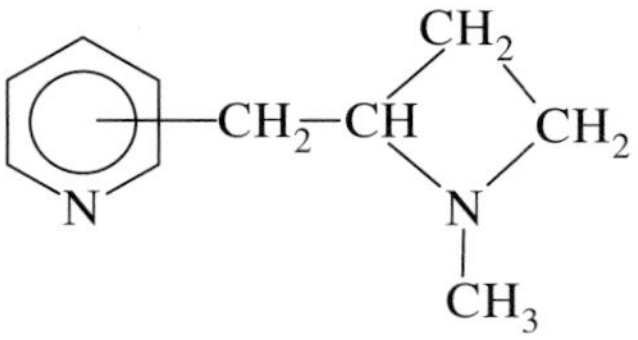

II

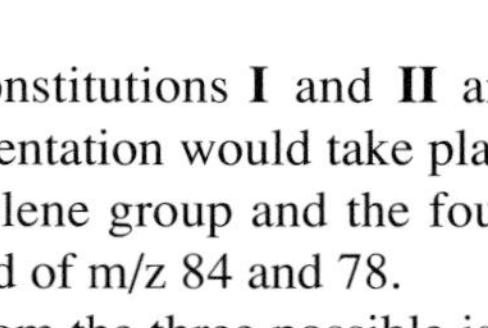

III

Constitutions **I** and **II** are excluded at once. In the mass spectrum a prominent fragmentation would take place between the two methylene groups in **I** and between the methylene group and the four-membered ring in **II**, both leading to m/z 60 and/or 92 instead of m/z 84 and 78.

From the three possible isomers of **III**, 4-substitution can be excluded for symmetry reasons. Although the molecule would be chiral, a fast rotation around the bond between the two rings would lead to two pairs of isochronous carbons and protons (cf. Chapter

19.4). The predicted values for the proton-NMR shifts are close to the experimental ones for 3-substitution and deviate markedly for 2-substitution. Thus, the substituent position is most likely 3. The unknown at hand is thereby identical to nicotine:

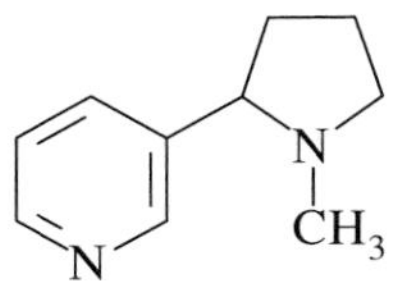

Check it in SpecTool:
To estimate proton chemical shifts of the pyridine ring select **Various Skeletons** from the **HNMR Tools** and then choose **Pyridine**. Click at a proton and select the substituent type from the then appearing menu, and then pick out a suitable substituent from the corresponding submenu. In the present case CH_2NH_2 is probably the best model for the 2-pyrrolidinyl substituent. Construct both isomers side by side and compare the predicted values. An analogous tool exists for carbon-13 NMR. The best available substituent is $CH_2CH_2CH_3$ in that case.

9.3 Comments

9.3.1 Mass Spectrum

The statement that an even mass fragment originating from an even mass molecular ion must be formed by H-rearrangement or otherwise contain an odd number of nitrogen atoms is slightly oversimplified, but implies a general rule: Lower mass fragments formed by a single bond cleavage in a radical cation are even electron ions of even mass only if they contain an odd number of atoms whose atomic weights and number of valencies are not both even or both odd. Since nitrogen is the only element in common organic compounds with an even atomic weight (14) and an odd number of valencies (3), the rule is usually applied as above. It assumes that no heavy isotopes like carbon-13, nitrogen-15, oxygen-17, deuterium etc. and no unusual elements such as Fe(III) and the like are present. Reactions involving multiple bond cleavages do, of course, require adaptation of the rule, e.g., retro-Diels-Alder reactions (equivalent to two single bond cleavages) yield even mass fragments without nitrogen being present, and double H-rearrangements yield accordingly odd mass products from even mass educts.

Loss of ethyl radical in nicotine (formation of the prominent m/z 133) is a common reaction in saturated alicyclic amines. The same holds true for the loss of hydrogen radical to form $[M-H]^+$.

Consult SpecTool:
Consult SpecTool for the general fragmentation scheme for cyclic amines.

9.3.2 Infrared Spectrum

Methyl groups and often also methylene groups bonded to a nitrogen atom in an amine exhibit characteristically low C–H stretching frequencies at or even below 2850 cm^{-1}. The present spectrum shows this very nicely. There are, however, other structural elements that also exhibit C–H stretching frequencies below 2850 cm^{-1}.

The band at 2500 cm^{-1} can be assigned to the str NH^+ vibration of the protonated nicotine which is formed from the traces of HCl always present in the solvent chloroform. The small band at 3650 cm^{-1} together with the bumpy baseline from 3400 - 3200 cm^{-1} indicates traces of water.

9.3.3 Proton and Carbon-13 NMR Spectra

An unambiguous assignment of the signals in the carbon-13 NMR spectrum is possible on the basis of additivity rules:

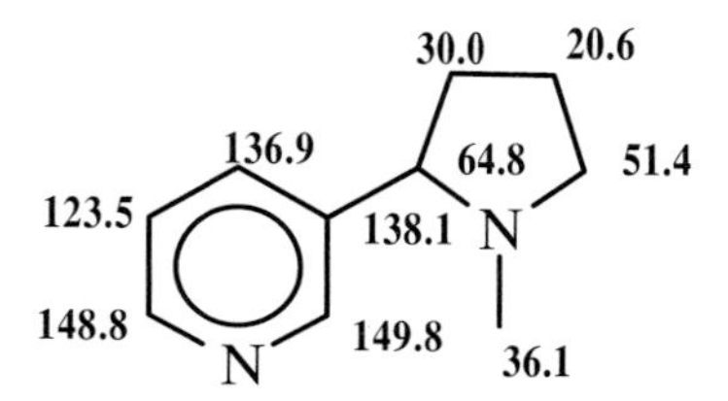

Consult SpecTool:
The data given above have been obtained with the general **^{13}C Shift Estimation** tool available in the **C13Tool** section. Use this program to predict the chemical shifts of the 2-substituted isomer and rationalize the differences. A more detailed description of the shift estimation tool can be found in the on page 22.

Assignment of the ^{1}H-NMR signals is not straightforward. All seven protons of the pyrrolidine ring exhibit different chemical shifts because the molecule is chiral and, therefore, the geminal methylene protons are diastereotopic. Assuming first order spectra, one would expect the methine proton at lowest field as a doublet of doublets, or a triplet if both vicinal coupling constants happen to be equal. The two methylene protons vicinal to the nitrogen should follow and exhibit 8 lines (doublet × doublet × doublet). Again, depending on the values of the coupling constants the number of lines may be reduced to anything between 7 and 4. Less than four lines can only be obtained if at least one coupling constant is small relative to the line widths. Surprisingly however, we find two signals at low field instead of one and two triplets (one of them with small further splittings) and one quartet instead of the expected minimal multiplicities of one triplet and two quartets. This clearly requires an explanation.

Consult SpecTool:
Access the ^{1}H-NMR spectrum of this compound by selecting **HNMR SpecLib** and Problem 9 and expand the relevant signals. You will notice that the signal at 3.23 ppm is not a triplet. As it has more than four lines it must be one of the methylene

protons. The pattern shows that it has two couplings which are very similar and one which is much smaller. The other methylene proton at 2.31 ppm, however, shows four sharp lines indicating three coupling constants (one geminal and two vicinal) of approximately equal size. Thus the signal at 3.09 ppm has to be assigned to the methine proton which also has two nearly equal couplings. A three-dimensional model (see Fig. 9.6) shows that one of the NCH_2 protons has a vicinal neighbor with a dihedral angle of close to 100°. According to the Karplus relation this means that the corresponding vicinal coupling constant must be much smaller than the other one. The two NCH_2 protons are in a very different position relative to the NCH_3 group and to the pyridine ring. This explains the large chemical shift difference of almost 1 ppm.

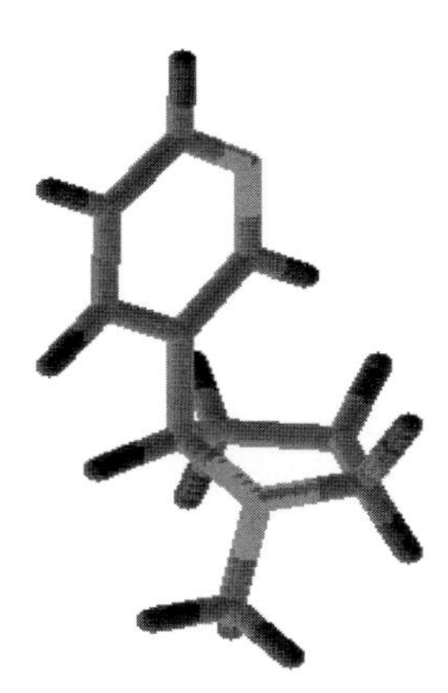

Fig. 9.6: Three-dimensional model of nicotine generated by the rule-based program ALCOGEN [1]

Expand also the aromatic signals and verify the assignment using multiplicities and coupling constants. Set the cursor over a line and then press the SHIFT key. If you now move the cursor to another line, the splitting is shown in Hz units at the top of the spectrum window.

9.4 References

[1] J. Sadowski, J. Gasteiger, Chem. Rev. 93 (1993) 2567.

10 Problem 10

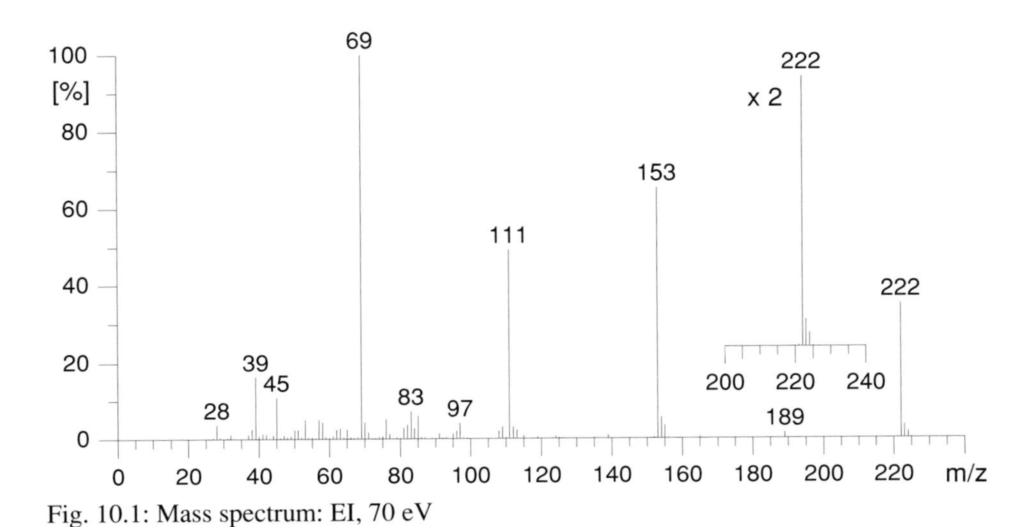

Fig. 10.1: Mass spectrum: EI, 70 eV

Fig. 10.2: IR spectrum: recorded in $CHCl_3$, cell thickness 0.2 mm

UV spectrum (solvent: ethanol):

λ_{max}	log ε
265	3.96
290	3.88
332	3.63

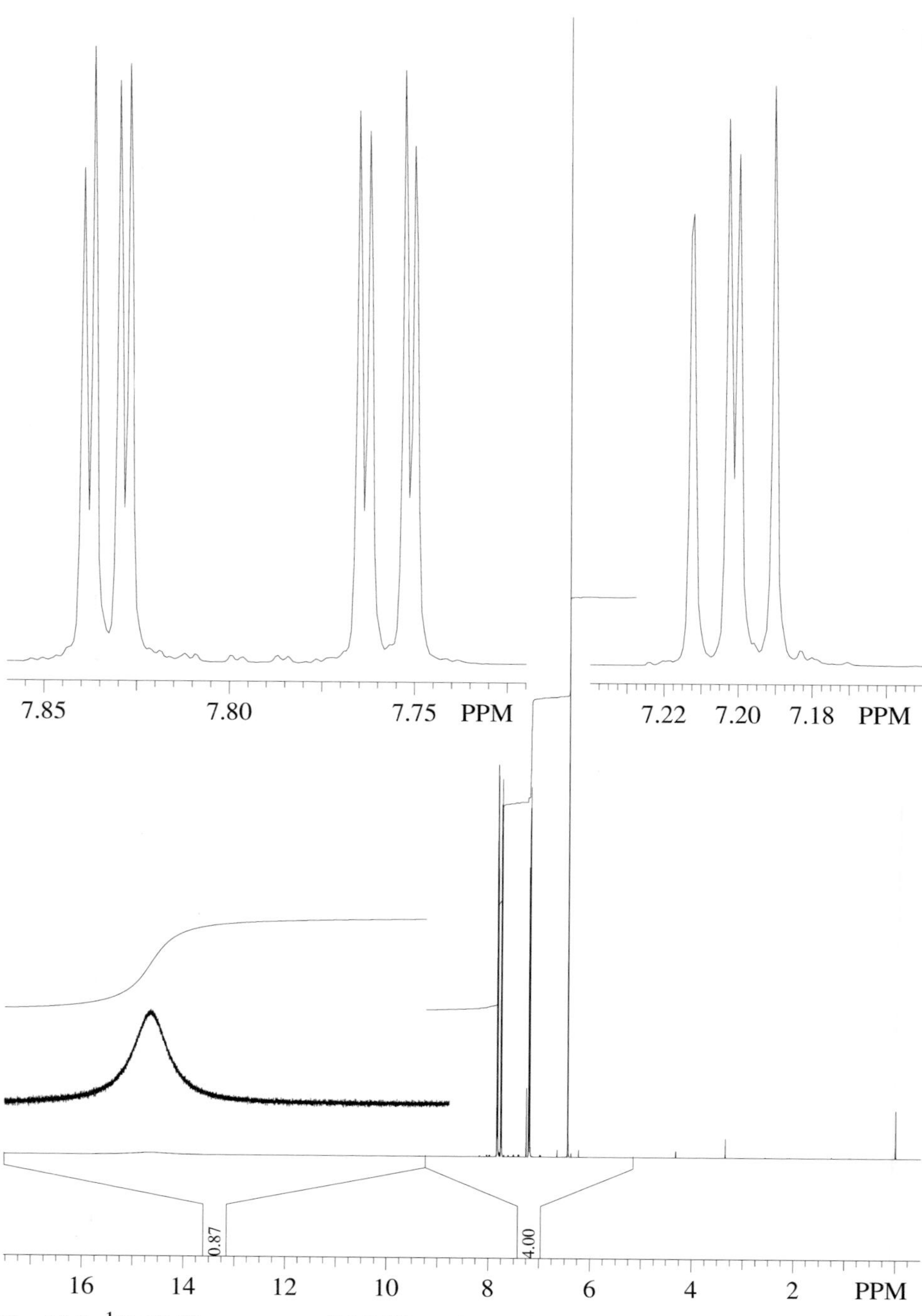

Fig. 10.3: ^{1}H-NMR spectrum: 400 MHz, solvent: $CDCl_3$

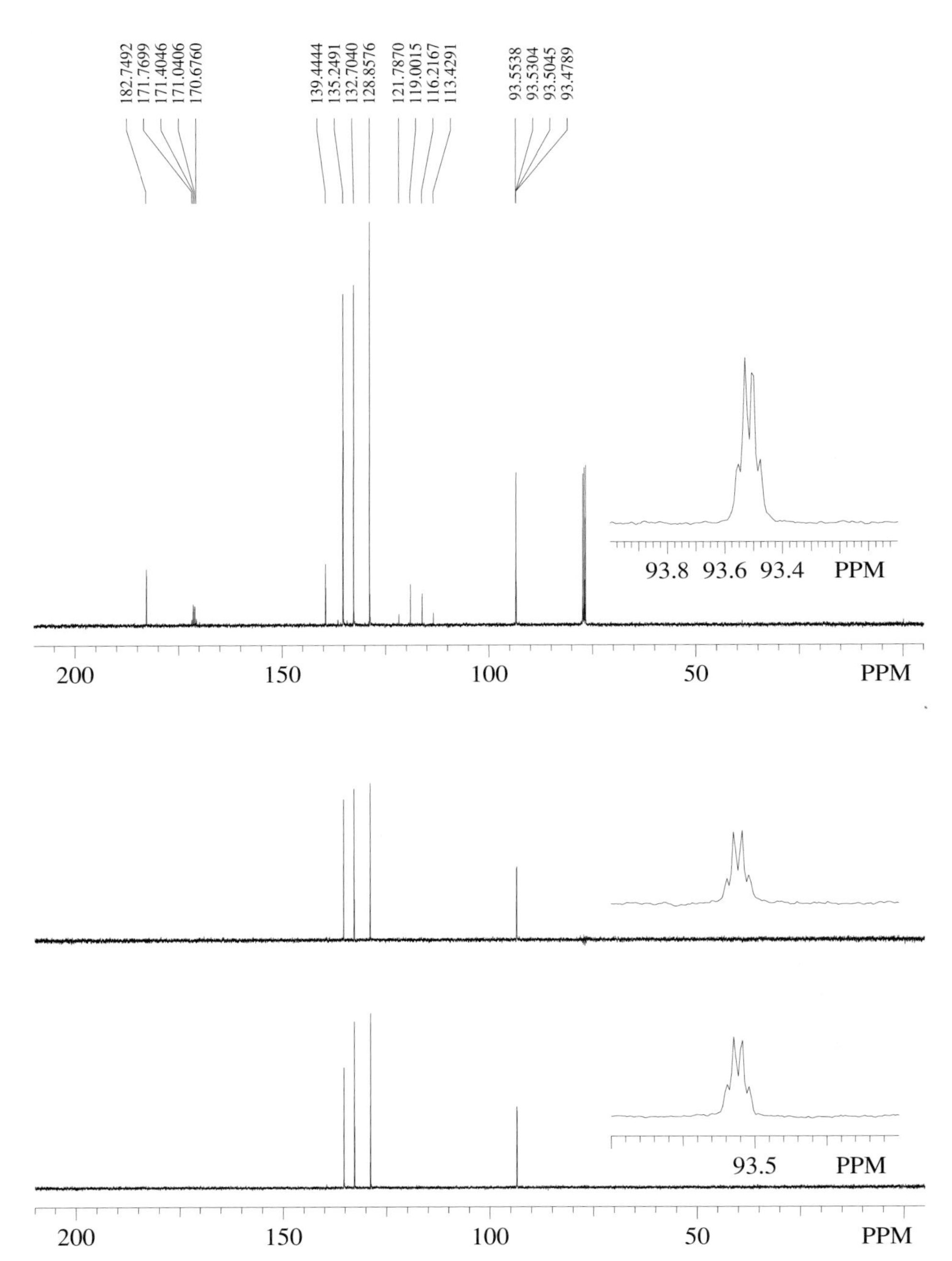

Fig. 10.4: ^{13}C-NMR spectrum: 100 MHz, solvent $CDCl_3$, Top: proton decoupled, middle: DEPT135, bottom DEPT90 ($\tau = 3.6$ ms)

10.1 Elemental Composition and Structural Features

The integration of the proton NMR spectrum leads to the intensity ratios of ca. 1 : 2 : 1 : 1. The chemical shift of δ = ca. 14.7 of the broad signal suggests the presence of a strongly hydrogen bonded hydroxyl group. The multiplets at 7 to 8 ppm establish an *AMX* spin system. There are no aliphatic protons in the molecule.

The carbon-13 NMR spectrum consists of 14 lines, four of which belong to CH groups according to the DEPT spectra. The presence of four methine groups and one proton not bonded to a carbon atom is in full accordance with the proton NMR spectrum.

The infrared spectrum shows a very broad absorption in the range of 2000 cm^{-1} to 3500 cm^{-1}, confirming the presence of a strongly hydrogen bonded hydroxyl group. There are very intense absorptions at 1590 cm^{-1} to 1650 cm^{-1} and at 1100 cm^{-1} to 1300 cm^{-1}, which are not easily assigned at this stage.

The mass spectrum ends with an intensive peak at m/z 222. There is no reason to assume that this peak does not correspond to the molecular ion. The intensity of the isotope signal at m/z 224 (ca. 5%) shows that the molecule contains one sulfur atom. Silicon would also lead to a signal of similar intensity at $[M + 2]^+$, but at the same time it would contribute to $[M+1]^+$ with 5%.

Check it in SpecTool:
Access **Isotope Table**, in the **MS Tools** section, find Si and S and then compare the natural isotope distribution of these elements.

The relative intensity of $[M + 1]^+$ is, however, only 10%, indicating that silicon is not present and that the molecule cannot contain more than nine carbon atoms. This fact is in conflict with the findings in the carbon NMR spectrum which contains 14 lines. The too large number of signals in the carbon-13 NMR spectrum could be explained by the presence of magnetic nuclei other than protons which lead to splittings of the carbon signals due to spin coupling. One hint for the presence of such nuclei is given by the mass spectrum.

Check it in SpecTool:
Access **MS Ranges** and try to find hints from the most intense peaks of the spectrum.

The two most intense peaks are m/z 69 and m/z 153 (M^+ - 69). If there are no aliphatic protons in the molecule, as is the present case, these fragments are characteristic of a trifluoromethyl group.

Consult SpecTool:
Confirm the presence of a CF_3 group on the basis of the carbon-13 NMR spectrum. You find references under **C-NMR Data**, **Halogen Compounds**, **Saturated Aliphatic**.

The carbon-13 chemical shift of a trifluoromethyl group is generally in the range of δ = 100 to δ = 120 and the C-F coupling constant is ca. 300 Hz. Since there are three magnetically equivalent coupling partners we expect a quartet. We indeed find four lines (δ = 121.8, 119.0, 116.2, 113.4) with a spacing of δ = 2.8 corresponding to 280 Hz in a 100 MHz spectrum. A closer inspection of the spectrum shows another quartet at ca. 171 ppm. This signal can be assigned to the carbon atom next to the trifluoromethyl group. According to the DEPT spectra it is a non-protonated carbon atom. The 14 lines in the carbon-13 NMR spectrum correspond thus to eight carbon atoms. The fragments found so far sum up to a molecular formula of $C_8H_5F_3OS$ accounting for a mass of 206 Daltons.

10.2 Structural Assembly

The following structural elements have now been detected:

Structural fragment	Mass
CF_3–C	81
CH–CH–CH	39
CH	13
OH	17
S	32
2C	24
Total mass	206

The difference of 16 mass units relative to the molecular mass of 222 Daltons suggests the presence of a further oxygen atom. The molecular formula is thus $C_8H_5F_3O_2S$ and corresponds to five units of unsaturation (if a divalent sulfur atom is assumed).

Consult SpecTool:
Calculate the expected isotope distribution with the MS Tool **Isotope Pattern** and compare it with the measured spectrum. Remember that you can zoom selected spectral regions in the **SpecLib** section which contains all spectra of this volume.

One of the carbon atoms is sp^3 hybridized (CF_3). With the remaining seven carbon atoms, one oxygen atom, and one sulfur atom we may construct at most four double bonds. Since allenes and acetylenes can be excluded on the basis of the carbon-13 chemical shifts (both would have two lines below δ = 90), the compound must have at least one ring. This ring has to contain the CH–CH–CH moiety. A benzene ring can be excluded, since we have only four signals for sp^2 hybridized carbon atoms in the aromatic region. Thus we have a five-membered heteroaromatic ring which may either be furan or thiophene.

Consult SpecTool:
Try to distinguish between furan and thiophene on the basis of the NMR spectra. Information is available under **CNMR Data** and **HNMR Data**, **Heteroaromatic**, **Five-membered**. Use the **Switch** function (**Menu** or **Palette**) to jump between the two methods.

Furan is less probable on the basis of the proton chemical shifts (a stronger shielding of all protons would be expected) and proton coupling constants (there is only one coupling constant larger than 3 Hz in furan and we observe two such couplings for the signal at $\delta = 7.3$) as well as on the basis of the carbon-13 chemical shifts (a stronger shielding occurs in furan). Also the rather intensive fragment at m/z 111 in the mass spectrum (with a sulfur isotope signal at m/z 113) perfectly fits to a thiophene derivative.

With the remaining elements we have to build a structure which allows a strong hydrogen bond of the OH proton. There are two such possibilities:

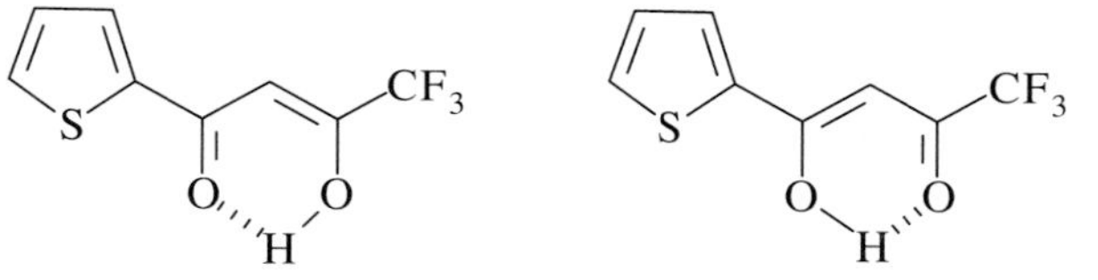

I **II**

The substituent must be attached to the thiophene ring in 2-position as indicated by the signal at $\delta = 7.3$ in the proton NMR spectrum which shows two couplings above 3 Hz. For the 3-substituted thiophene only one such coupling of the high-field proton would occur.

Check it in SpecTool:
Use the **C13 Shift Estimation** tool to predict the ^{13}C-NMR spectrum for both tautomers **I** and **II**.

On the basis of the data presented here tautomer **I** seems to be predominant.

10.3 Comments

10.3.1 Mass Spectrum

The main features of the spectrum seem to arise by simple bond cleavages next to the carbonyl groups. The trifluoromethyl group is lost to give m/z 153 and appears as charged fragment at m/z 69, probably in part by decarbonylation from the acyl ion m/z 97. The peak at m/z 111, which could be either half of the molecule, seems to be entirely due to thiophenoyl ion as judged from the ^{34}S isotope peak at m/z 113. Both the appearance of m/z 189 (loss of SH from $M^{+\cdot}$) and formation of CHS^+ (m/z 45) could be taken as additional indications of the presence of sulphur. There is no indication of enolization in gas phase.

10.3.2 Infrared Spectrum

Intermolecular hydrogen bonded hydroxyl groups generally give rise to a moderately sharp band between 3600 and 3450 cm^{-1}. If more than one hydrogen bond is involved with a particular hydroxyl group, the band becomes broader and moves to lower frequencies (3400 to 3200 cm^{-1}). Intermolecular as opposed to intramolecular hydrogen bonding is characterized by the band shape and frequencies being highly dependent on the sample concentration. Intramolecular hydrogen bonding, on the other hand, is not sensitive to sample concentration. It may, however, be influenced by the solvent. In the general case, a band similar to the one observed with intermolecular hydrogen bonds is found. However, if extremely strong hydrogen bonds are formed as for example in enolized β-diketones or in *ortho* nitrophenols, very broad bands extending from 3200 cm^{-1} down to 2000 cm^{-1} result. As the total intensity is then spread over a wide range, such bands may easily be overlooked.

Carbonyl stretching frequencies are lowered by hydrogen bonding. If very strong intramolecular hydrogen bonds are formed as, e.g., in enolized β-diketones, the stretching frequencies of the two CO and the two CC bonds in the enol molecule become similar and strong interactions occur. Therefore, it is no longer possible to assign the resulting broad absorption at about 1600 cm^{-1} to either the hydrogen-bonded carbonyl or the C=C double bond. The absorption must rather be considered as being due to the whole conjugated chromophore. Another group of strong bands arising predominantly from C–O stretching vibrations is observed around 1250 cm^{-1}.

The carbon-fluorine stretching frequencies fall into the range from 1400 cm^{-1} to 1000 cm^{-1}. Due to rotational isomerism, several sharp bands, sometimes fused to a broad strong band are observed. As other chromophores give rise to strong absorptions in this region, fluorine is difficult to identify from the infrared spectrum.

10.3.3 Ultraviolet Spectrum

The ultraviolet spectrum of this compound can hardly be rationalized without extensive calculations. However, it fits the general picture of thiophenes substituted in 2-position with a carbonyl group. For such compound at least two bands are expected, separated by some 20 nm. Both bands have an extinction coefficient around 10 000, the lower wavelength band generally being slightly more intense.

11 Problem 11

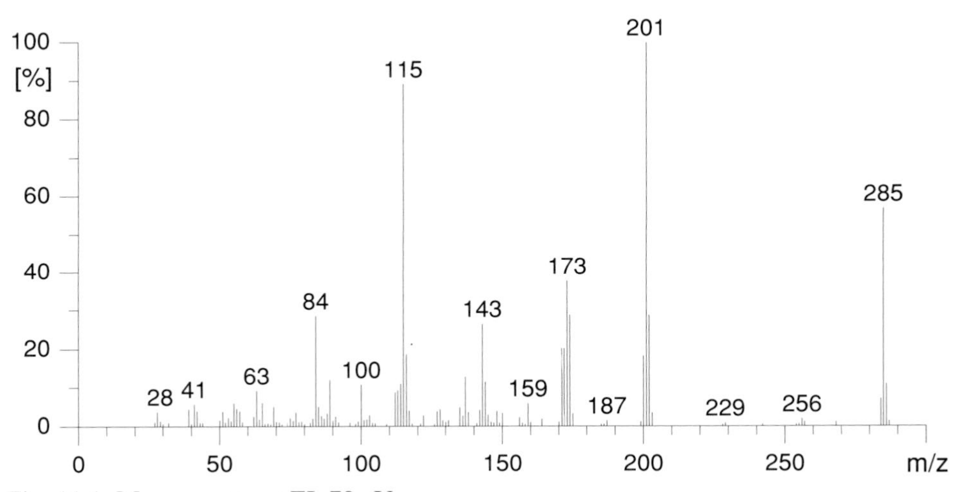

Fig. 11.1: Mass spectrum: EI, 70 eV

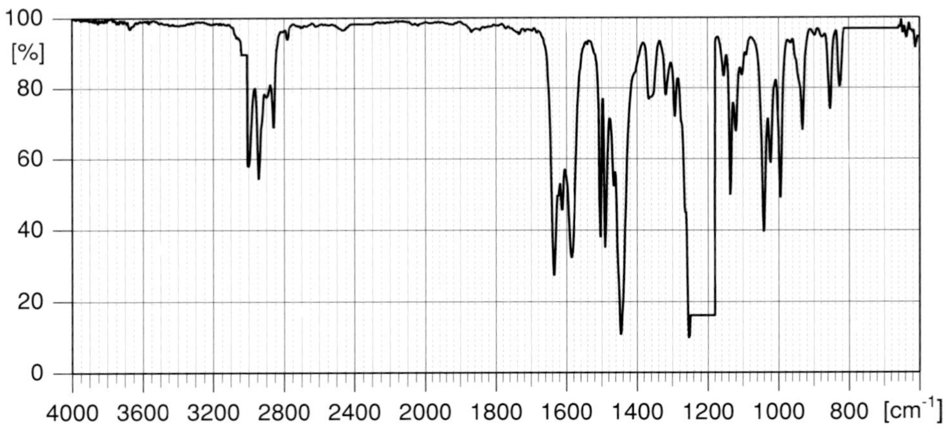

Fig. 11.2: IR spectrum: recorded in $CHCl_3$, cell thickness 0.2 mm

UV spectrum (solvent: ethanol):

λ_{max}	log ε
244	4.37
258	4.37
310	4.64
343	4.84

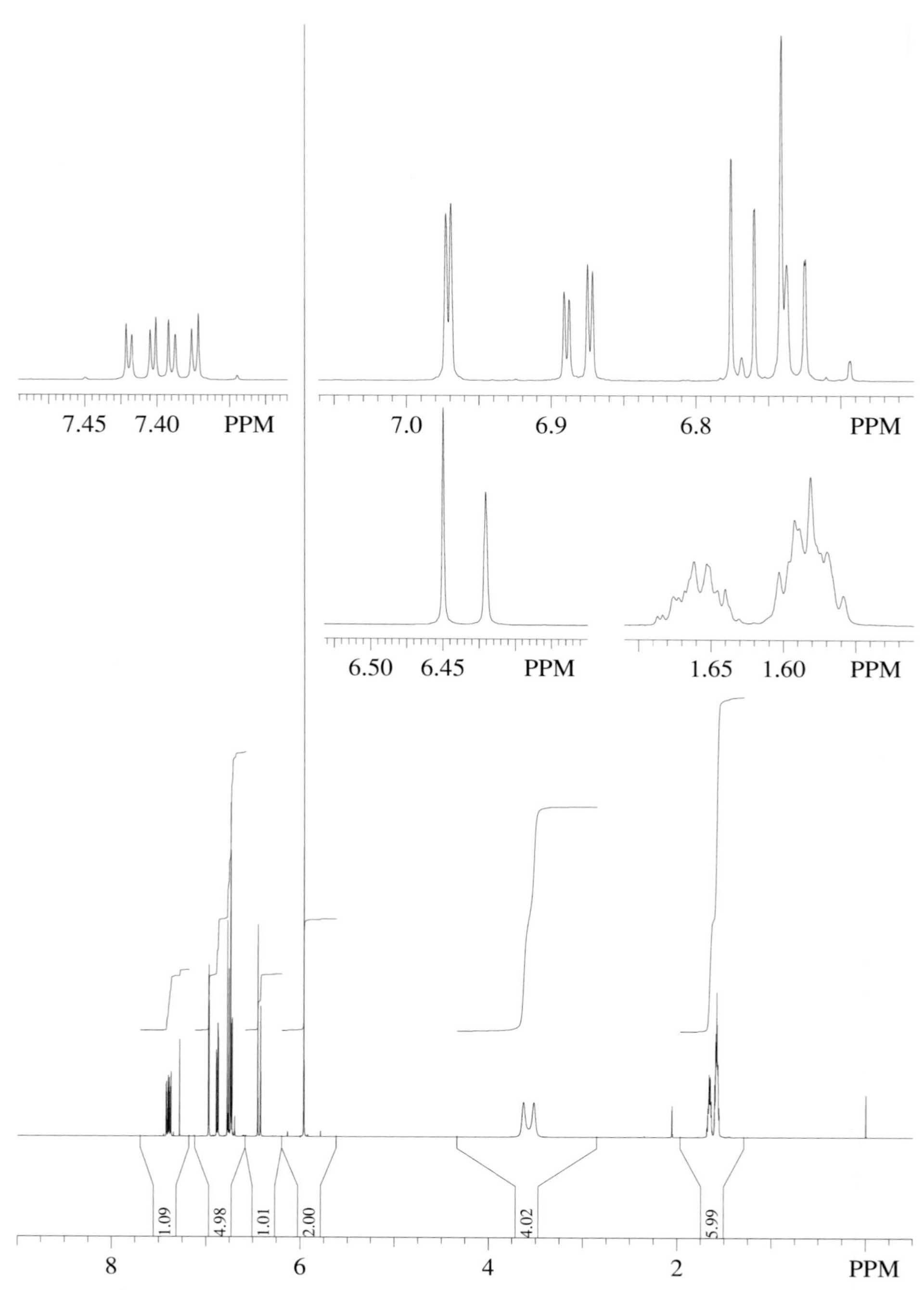

Fig. 11.3: ^{1}H-NMR spectrum: 500 MHz, solvent: $CDCl_3$

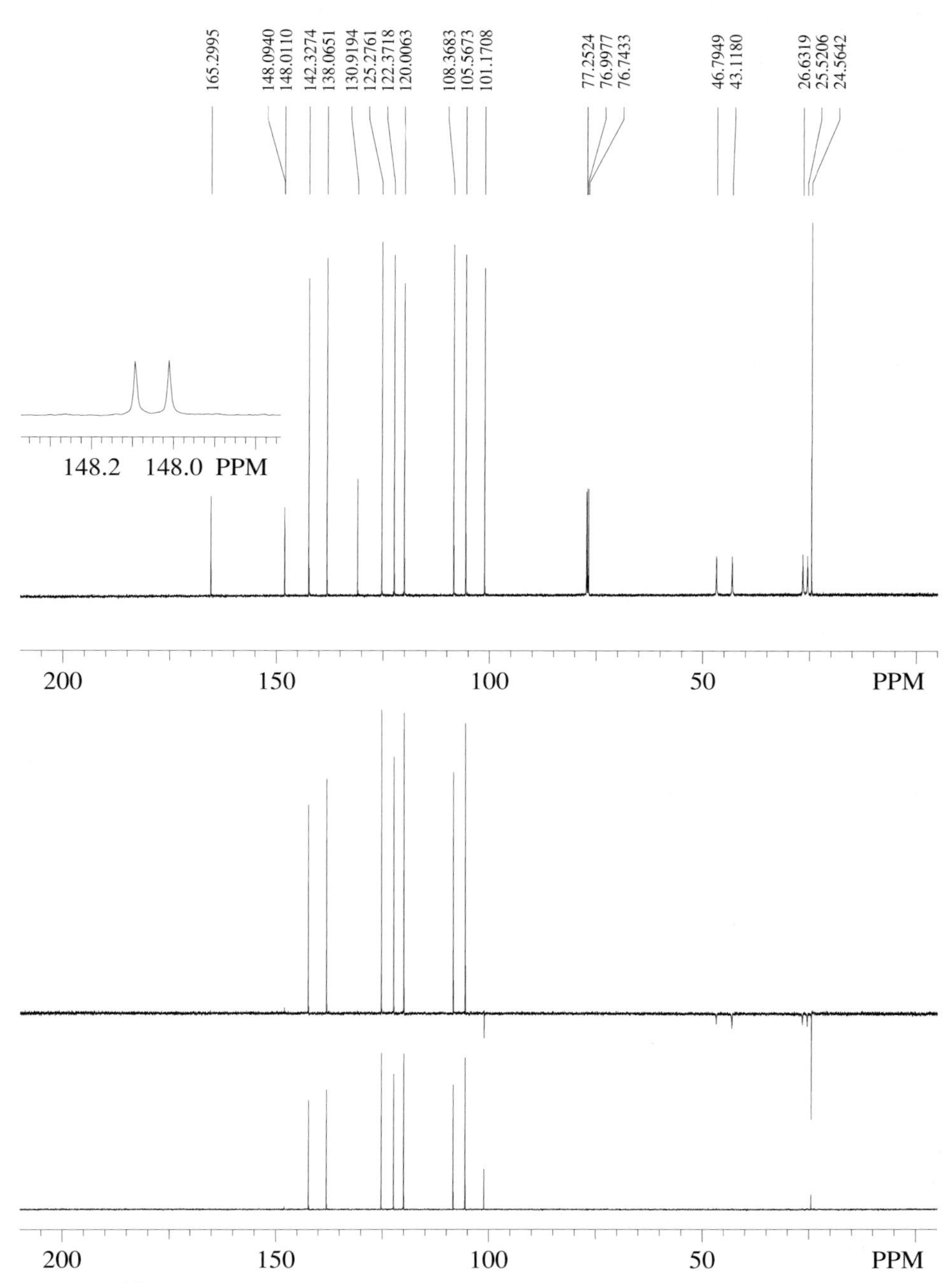

Fig. 11.4: ^{13}C-NMR spectrum: 125 MHz, solvent $CDCl_3$, Top: proton decoupled, middle: DEPT135, bottom DEPT90 ($\tau = 3.6$ ms)

11.1 Elemental Composition and Structural Features

In the infrared spectrum, we note CH stretching frequencies above 3000 cm^{-1} as well as at the lower end of the standard range at 2850 cm^{-1}. The former indicates hydrogen atoms bonded to sp^2 hybridized carbon atoms, whereas the latter suggests methyl or methylene groups bonded to the heteroatom in an ether or an amine. The strong bands at 1640 cm^{-1} and 1580 cm^{-1} could be due to carbonyl groups which have to be strongly delocalized in order to explain their low stretching frequencies. The doublet at 1500 cm^{-1} strongly suggests a benzene ring.

The mass spectrum ends with a peak cluster centered at m/z 285. This value may be taken as the molecular mass, as all differences to lower mass peaks are chemically reasonable. 285 being an odd number indicates an odd number of nitrogen atoms. The general type of the spectrum is both aromatic (intensity concentrated at a few peaks in the higher mass range), and nonaromatic according to mass values of the fragments in the lower mass range. We note that the molecular ion predominantly loses 84 mass units, the same fragment also forms the most important peak in the low mass region.

The proton NMR spectrum yields upon integration a proton ratio of (from high to low shift values) 1 : 1 : 1 : 3 : 1 : 2 : 4 : 6, summing up to a total of 19 protons. From the integral of the lines at $\delta = 3.6$ and $\delta = 1.6$ we assume that five methylene groups are present, two of them bonded to a heteroatom. This heteroaliphatic moiety might correspond to the fragment of mass 84 which is lost from the molecular ion in the mass spectrum, leaving a predominantly aromatic remainder. With five methylene groups we need 14 mass units to complete the fragment. The nitrogen atom already inferred from the mass spectrum nicely fits in here. We thus conclude the presence of a piperidine ring.

Check it in SpecTool:
Access the **MS**, **CNMR** and **HNMR** data and verify the hypothesis of the presence of a piperidine ring. You find saturated cyclic amines under sp^3 hybridized nitrogen.

Another conspicuous signal is the very sharp singlet at $\delta = 5.95$. A sharp signal corresponding to two protons with this chemical shift value is rather uncommon. Very often, it is due to a methylenedioxy group on an aromatic ring. To check for this possibility we search the carbon-13 NMR DEPT135 spectrum for a CH_2 at about $\delta =$ 100, which is the expected chemical shift value for the carbon atom in a methylenedioxy group. We indeed find such a signal, which confirms our assumption (cf. Comments on DEPT spectra of Problem 7).

Check it in SpecTool:
Compare the observed chemical shifts of the methylenedioxy group with reference data given under **Acetals Ketals Ortho Esters** in the **Ether** section of **HNMR Data** and **CNMR Data**.

Furthermore, at δ = 6.44 we see one half of an *AB* spin system with a coupling constant of J = 16 Hz. This is indicative of a *trans* disubstituted carbon-carbon double bond.

Check it in SpecTool:
Overview of the various coupling constants can be found under **Special** in the **HNMR Data** section. You also find data on coupling constants for double bonds under **Alkene**, $\mathbf{J_{gem}}$, $\mathbf{J_{vic}}$.

The carbon-13 NMR spectrum shows a signal at δ = 165.3 which we assign to a carbonyl group already inferred from infrared spectral evidence. As there is only one signal, we will have either to explain why one carbonyl group gives rise to several bands around 1600 cm^{-1} in the infrared spectrum, or we have to assume two different carbonyl groups of identical chemical shift values in the carbon-13 NMR spectrum. Additionally, we count eleven signals in the region where sp^2 hybridized carbon atoms usually resonate. One has already been assigned to the methylenedioxy group. We have thus, in addititon to the carbonyl group, ten sp^2 hybridized carbon atoms, three of them have no attached hydrogens according to the DEPT spectra, the other seven being methines. The two carbon atoms at δ = 148.0 and δ = 148.1 are assigned to the two benzene carbon atoms bearing the methylenedioxy group.

Check it in SpecTool:
Reaccess **Acetals Ketals Ortho Esters** page in the **Ether** section of **CNMR Data** to verify this assignment.

The third quarternary carbon has also to be assigned to a benzene carbon, as otherwise we would end up with two separate molecules.

The remaining signals have to be assigned to the piperidine ring. If the piperidine ring was free to rotate we would find just three sharp signals. On the other hand, if it was in a fixed conformation, five sharp signals would be expected in absence of symmetry. With intermediate rates of rotation, we will find broadened lines, as observed here. Thus, the piperidine ring is in an environment where it rotates with intermediate speed as measured on the time scale defined by the reciprocal difference of the chemical shift values of the exchanging nuclei (measured in Hertz) at the given magnetic field strength. Line broadening in nitrogen compounds is most commonly observed in amides (slow rotation about the peptide bond). We, therefore, may tentatively postulate that the piperidine nitrogen is substituted with a carbonyl group. If we further assume that this carbonyl group is in conjugation with a double bond, its infrared stretching frequency as well as its carbon-13 NMR chemical shift can be rationalized.

We may now summarize the fragments identified so far:

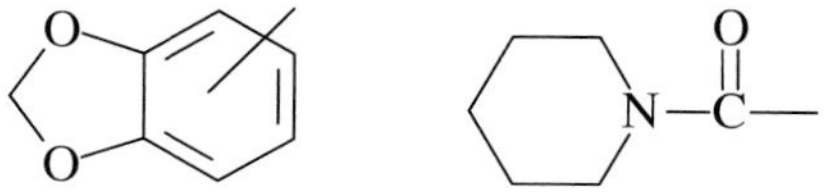

O O N—C(=O)— C_4H_4

The elemental composition is calculated to be $C_{17}H_{14}NO_3$ with a molecular mass of 285. This is in perfect agreement with the value indicated by the mass spectrum. Our molecule is thus complete.

11.2 Structural Assembly

As the C_4H_4 moiety consists of four methine groups (cf. carbon-13 DEPT NMR spectra) there is but one way of assembling the structural fragments, namely:

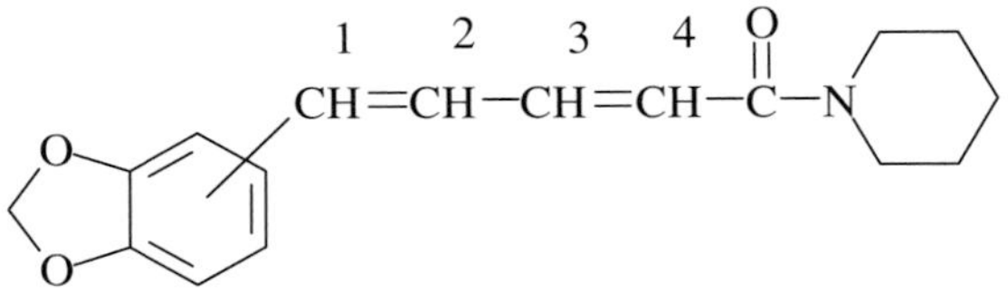

1 2 3 4
CH=CH–CH=CH–C(=O)–N

However, the substituent position on the benzene ring as well as the stereochemistry of the two double bonds remains to be determined. We already know that at least one of the double bonds is in the *trans* form. We have thus to consider only the *trans-trans* and both *cis-trans* forms. We proceed by estimating the proton NMR chemical shift values.

Check it in SpecTool:
Use the **HShift** Tool to estimate the chemical shifts for the three possible configurations.

We may exclude at once the *trans-cis* isomer as it does not provide for the signal at $\delta = 7.4$. However, the chemical shift estimation does not distinguish between the *cis-trans* and *trans-trans* forms. In both compounds, the signal at 7.4 ppm can be unambiguously assigned to the proton in 3 position (*cis* to the carbonyl group). On closer inspection of the respective pattern one may see that it indeed contains a large coupling constant, indicating *trans* substitution. The proton at 6.4 ppm clearly has only one vicinal coupling partner so that it must be either at 1 or at 4 position. However, since the remaining two protons at ca. 6.72 and 6.75 ppm are strongly coupled to each other, they are on the same double bond, i.e., they must be assigned to H-1 and H-2. Despite the presence of a higher order spin system, the separation of the two rightmost lines in the multiplet (16 Hz) indicates that the system includes one *trans* coupling. Therefore, both double bonds are *trans* disubstituted.

The three aromatic protons can be easily assigned to the signals at ca. 6.97, 6.88 and 6.78 ppm. They exhibit one small, one large, and one small and one large coupling, respectively. Such a pattern clearly indicates a 1,2,4-trisubstitution. Thus the compound is:

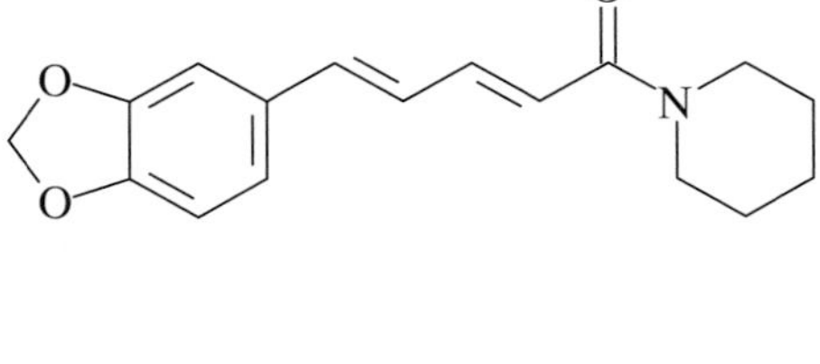

11.3 Comments

11.3.1 Mass Spectrum

Fragmentation of the compound follows the expected path. Initial amide cleavage yields m/z 201 and 84, the major aromatic product eliminates carbon monoxide (28) and formaldehyde (30) in succession down to the unsaturated hydrocarbon residue m/z 115 ($C_9H_7^+$) which further dehydrogenates. Amide cleavage is accompanied by a hydrogen transfer reaction (possibly of McLafferty type with the double bond next carbonyl as hydrogen acceptor) to produce m/z 202, which also loses carbon monoxide and formaldehyde. This interpretation has been confirmed by independent experiments (not shown here).

11.3.2 Proton and Carbon-13 NMR Spectra

The interpretation of higher order spectra is delicate, often anti-intuitive and sometimes outright misleading. Therefore, confirmation by spectrum simulation is an absolute must. For example, for the proton at 7.4 ppm four lines would be expected in a first order spectrum since it has two coupling partners. The presence of 10 lines (including the two small ones) is a consequence of second order effects. The interpretation above has been confirmed by spectrum simulation which is treated in another volume of this series [1].

If there is an exchange between two or more environments of a nucleus, single lines are observed if the average life time of the system is short relative to the reciprocal chemical shift differences of the different environments (measured in Hertz), while discrete lines are observed if the average life time of the system is long, and broad lines are observed for intermediate exchange rates. The NMR time scale is defined by the chemical shift differences (in Hz) for the nuclei in different environments. Since the line frequencies depend on the magnetic field strength, the NMR time scales are also dependent on it. In contrast to the spectrum given here, in a proton NMR spectrum recorded at 100 MHz at the same sample temperature, the two lines of the piperidine methylene protons next to the nitrogen atoms at 3.6 ppm give rise to one single broad line because the rotation around the CO–N bond at that magnetic field strength is "fast". Even in the high-field ^1H- and ^{13}C-NMR spectra at hand, the exchange still leads to some line broadening.

11.4 References

[1] U. Weber, H. Thiele, G. Hägele, NMR Spectroscopy: Modern Spectral Analysis, VCH, Weinheim, 1997.

12 Problem 12

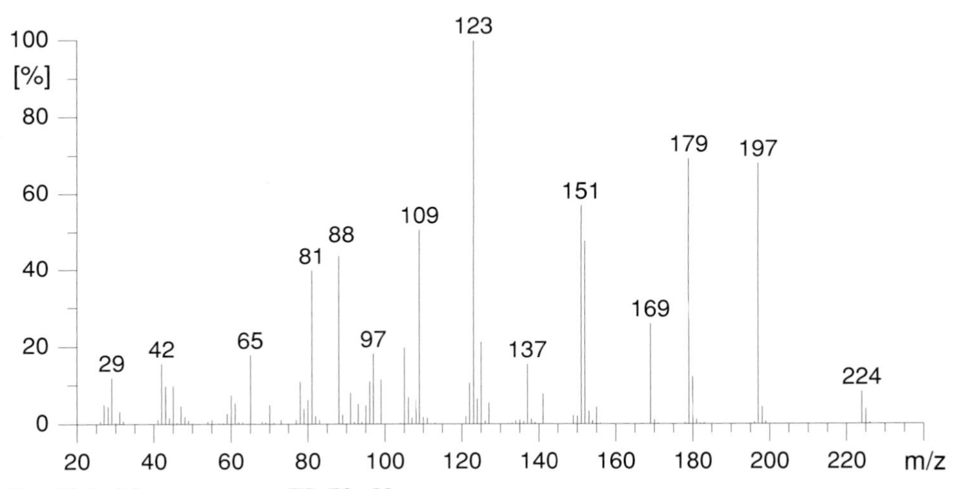

Fig. 12.1: Mass spectrum: EI, 70 eV

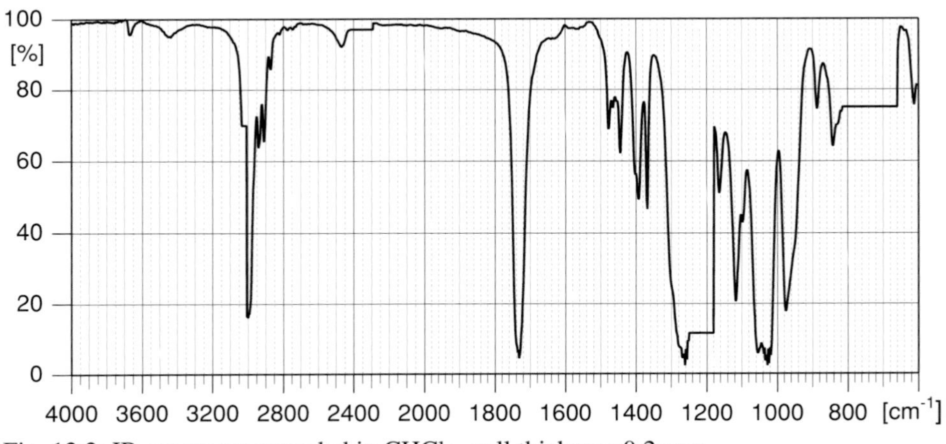

Fig. 12.2: IR spectrum: recorded in $CHCl_3$, cell thickness 0.2 mm

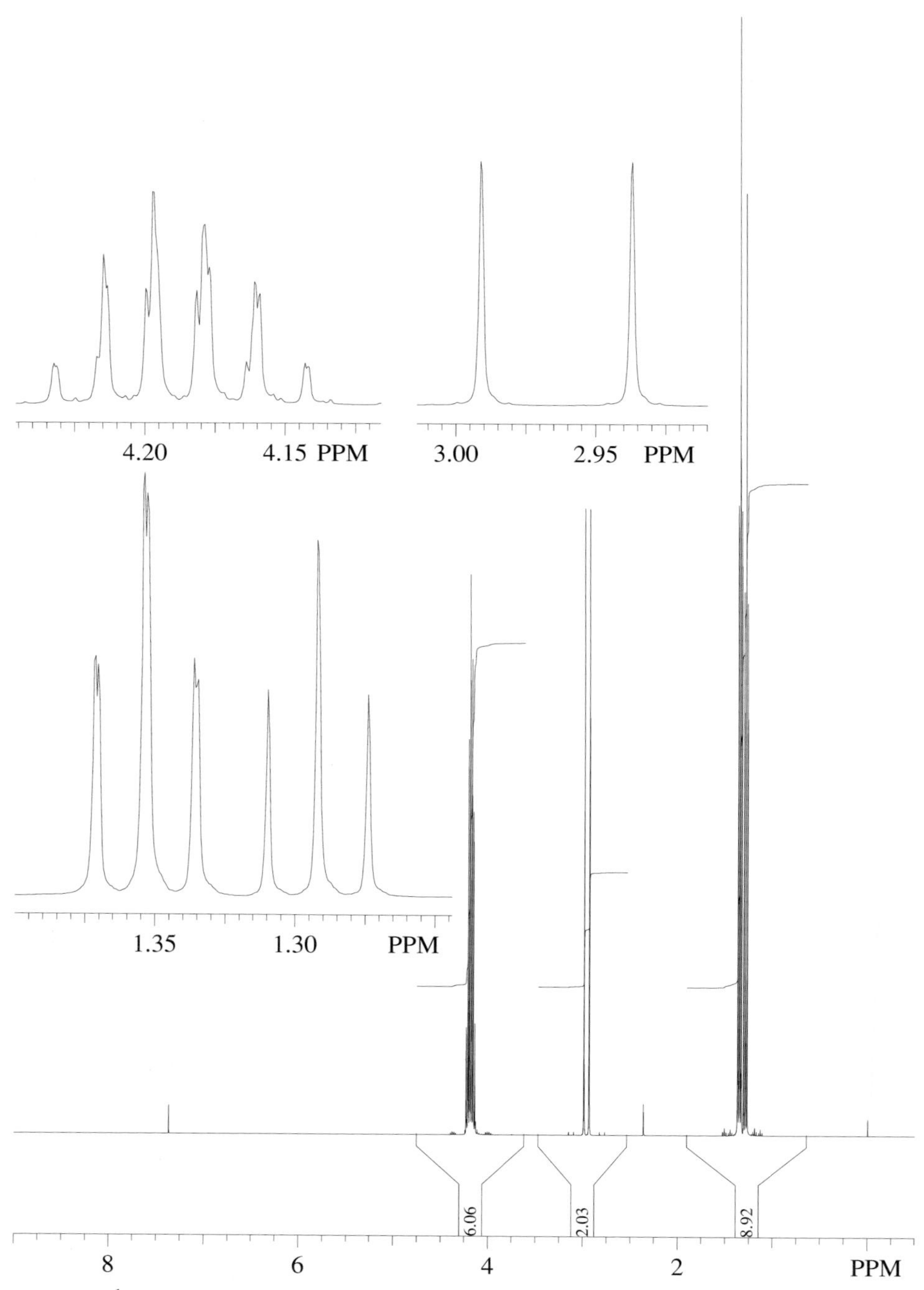

Fig. 12.3: ^{1}H-NMR spectrum: 400 MHz, solvent: $CDCl_3$

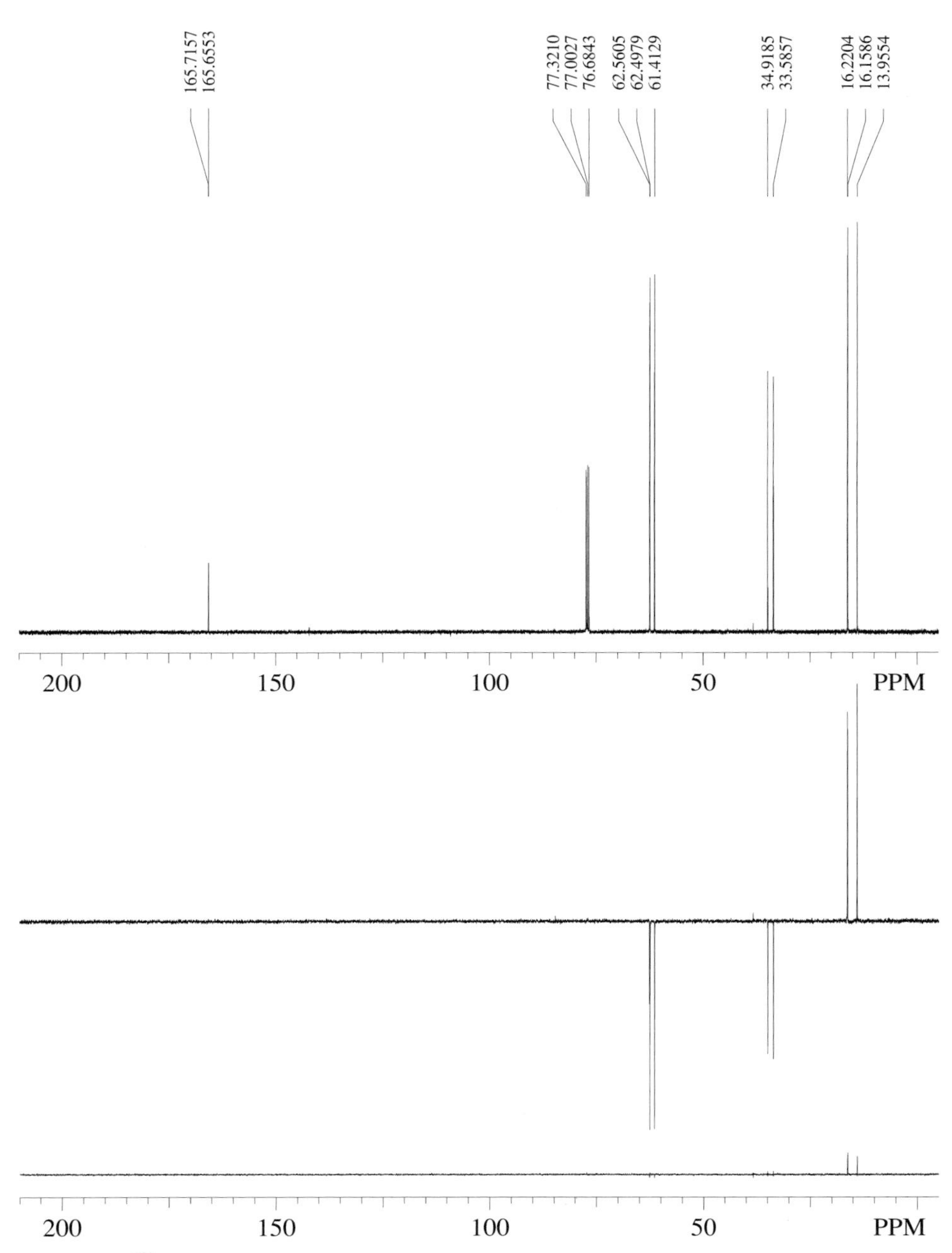

Fig. 12.4: ^{13}C-NMR spectrum: 100 MHz, solvent $CDCl_3$, Top: proton decoupled, middle: DEPT135, bottom DEPT90 ($\tau = 3.6$ ms)

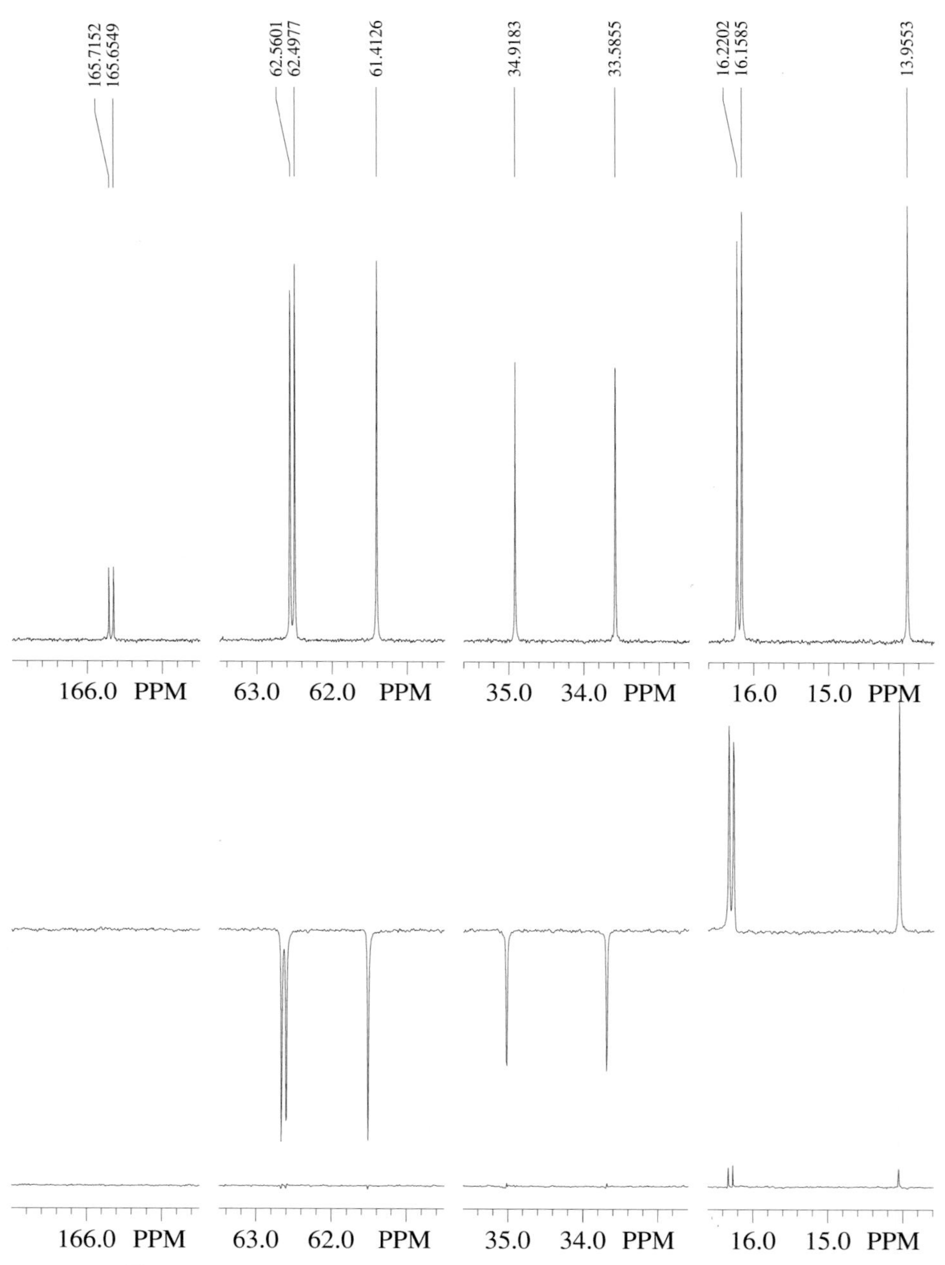

Fig. 12.5: ^{13}C-NMR spectrum (expanded regions): 100 MHz, solvent $CDCl_3$, Top: proton decoupled, middle: DEPT135, bottom DEPT90 (τ = 3.6 ms)

12.1 Elemental Composition and Structural Features

Mass spectrum: m/z 224 probably represents the molecular ion. The intensity distribution indicates a nonaromatic system. Loss of 45 mass units from $M^{+\cdot}$ (224 → 179) and fragments at m/z 31, 45, and 88 prove that oxygen is present.

Check it in SpecTool:
Access **Indicators for Heteroatoms** of the **MS Tools** section and use the button **Empirical Rules**.

The peak at m/z 197 is accompanied by a conspicuously small ^{13}C-isotope peak of only 7% relative intensity, which allows for not more than about six carbon atoms (see also Comments to Problem 8). The contribution of hetero atoms to the molecular formula must, therefore, be very large.

Infrared spectrum: Weak bands at around 3660 and 3450 cm^{-1} suggest hydroxyl stretching vibrations (free and associated, respectively), but taken together with the shoulder at 1650 cm^{-1} it may only indicate water present as an impurity. A strong band at 1735 cm^{-1} shows the presence of a carbonyl group. Broad bands of high intensity around 1260 and 1040 cm^{-1} could be attributed to B–O, C–N, C–F, S=O, C–O, P=O, P–O or C=S stretching vibrations, among which boron and sulfur containing functions are ruled out by the missing natural isotope contribution to m/z 197 (at m/z 196 or 199, respectively).

Check it in SpecTool:
Access **TOP**, **MS Tools**, **Isotope Table** to check natural isotope abundances.

Proton NMR spectrum: The integral ratios from left to right are 6 : 1 : 1 : 9, summing up to 17 protons. The high-field multiplet consists of two triplets for two and one methyl groups, respectively, which must be coupled to protons giving rise to the multiplet around 4.2 ppm. If the nine proton multiplet at high field is attributed to three methyl groups then the multiplet at low field must be due to three overlapping methylene quartets, constituting three ethoxyl groups according to the chemical shift value. The two remaining signals could represent either two isolated methine protons or one methylene group coupled to a hetero atom with J = 22 Hz.

Carbon-13 NMR spectrum: Ten signals are noted in the wide-band decoupled spectrum corresponding to three CH_3, five CH_2, and two C according to the DEPT spectra. This corresponds to 19 hydrogen atoms but the proton NMR spectrum shows only 17. The tree ethoxyl groups identified in the proton NMR spectrum are evident in the carbon-13 NMR spectrum as the three signals at 13-16 ppm (CH_3) and 61-63 ppm (CH_2). However, the two CH_2 signals at 33-35 ppm in the carbon-13 NMR correspond to only two protons at 2.9-3 ppm in the proton NMR spectrum. The most reasonable explanation for this discrepancy is the presence of a heteronuclear coupling of carbon-13 and of the protons with a nucleus of spin I = 1/2, resulting in doublets with coupling constants of 134 Hz and 22 Hz for the methylene group in the carbon-13 and proton NMR spectra, respectively. Phosphorus is the most likely candidate for such a nucleus, because its presence is suggested by the strong bands in the region 1250 to 1050 cm^{-1} in the infrared spectrum and by some features in the mass spectrum (a signal at m/z 47, the missing carbon-13 isotope peak for the fragment of m/z 97 and the isolated appearance of m/z 65).

> **Check it in SpecTool:**
> Access the page **TOP**, **IR Data**, **Phosphorus Compounds**, **P=O Compounds** and click on the band or on the panel to present reference data or comments, respectively. Verify the non-specific phosphorus indicators in the mass spectrum on the page **TOP**, **MS Data**, **Phosphorus Compounds**, **Phosphates**.

According to this interpretation, there is only one C=O and one CH_2 carbon atom in addition to the three ethoxyl groups. A mass balance for the structural features, necessary to account for the observed spectroscopic data, results in the following list:

$-OCH_2CH_3$	$-OCH_2CH_3$	$-OCH_2CH_3$	C=O	$-CH_2-$	P	$C_8H_{17}O_4P$
45	45	45	28	14	31	208

16 mass units are missing, (if the molecular mass is assumed to be 224, as suggested by the mass spectrum) and must be attributed to one additional oxygen atom, because no further protons are available. The molecular formula then becomes $C_8H_{17}O_5P$.

12.2 Structural Assembly

The carbonyl group is by its chemical shift value in the carbon-13 NMR spectrum and by its stretching frequency in the infrared spectrum an ester carbonyl and constitutes, considering the available structural elements, necessarily an ethyl ester moiety. Since the proton and carbon-13 chemical shift of the methylene group (2.97 and 34.2 ppm respectively) excludes a neighboring oxygen atom, the following constitution is the only chemically meaningful possibility:

$$(CH_3-CH_2-O)_2P(=O)-CH_2-C(=O)-O-CH_2-CH_3$$

12.3 Comments

12.3.1 Mass Spectrum

Loss of 27 mass units from the molecular ion to form a prominent peak (which has been shown by independent experiments to be the starting point to most of the subsequent degradations) arises by a double hydrogen rearrangement from ethoxyl which is typical for alkyl phosphates and rather common with many alkyl esters of carboxylic acids as well. The formation of the even number maximum at m/z 88 is the result of a McLafferty reaction which involves transfer of one of the rearranged hydrogen atoms from oxygen on to the ester carbonyl group in a six membered cyclic transition state and formation of an ethyl acetate ion. The fragment at m/z 65 could have been interpreted as indicating the presence of phosphorus oxide (PO_2H_2) by itself, because the absence of adjacent isotope peaks rules out the usual hydrocarbon composition C_5H_5. Alkyl phosphates yield an m/z 99 (PO_4H_4) of similar appearance.

A search for characteristic ion sequences in the lower mass range, which usually is one of the first measures taken in order to get some general information about the compound type from the mass spectrum, does not yield any sequence exceeding three members (m/z 31, 45, 59 in the oxygen series is the only one perceivable). In such cases the conclusion is appropriate that no coherent carbon skeleton of significant length is present, because consistent ion sequences depend on a carbon chain of some minimal extension. Especially noteworthy is the absence of the otherwise ubiquitous fragment m/z 39 ($C_3H_3^+$), which practically excludes a hydrocarbon backbone of three carbon atoms or more. A missing m/z 41 ($C_3H_5^+$) is of similar significance in a nonaromatic system.

12.3.2 Infrared Spectrum

In this comparatively simple compound, most bands in the fingerprint region between 1500 and 1000 cm^{-1} may be rationalized. At 1480 cm^{-1} we find the deformation vibration of the methylene groups in the ethoxyl substituents on the phosphorus atom, at around 1445 cm^{-1} is the asymmetric deformation vibration of the methyl groups. Between these two absorption bands we have the respective vibrations of the ethyl ester group. In the relatively broad band around 1400 cm^{-1} the absorptions due to deformations of the methylene group between the phosphorus atom and the carbonyl group (at around 1410 cm^{-1}), the wagging vibration of the other methylene groups (at 1395 cm^{-1}) overlap. The symmetric deformation vibration of the methyl groups in the ethoxyl groups on the phosphorus atom gives rise to the absorption at 1375 cm^{-1}. The C–O–P moiety gives rise to the strong absorption at 1050 cm^{-1}. Splitting of this band into a doublet is characteristic for ethoxyl groups on phosphorus. Such a splitting is observed neither in the methoxyl- nor in higher alkoxy phosphorus compounds.

12.3.3 Proton NMR Spectrum

The coupling constant $^{1}H–C–^{31}P$ across three bonds is around 8 Hz and exhibits thus nearly the same value as the vicinal proton-proton coupling constant. In a first order spectrum with isochronous methylene protons a quintet may occur for the methylene protons. In the present case, the geminal methylene protons of the two ethoxyl groups equivalent by symmetry are diastereotopic (see also Chapter 19.4) and the predicted spin-system is of A_3MNX type. In reality, the two systems must not be considered independently as they are interlinked by the phosphorus atom (X) as a common coupling partner so that the spin system is in fact $A_3A_3'MM'NN'X$. Its lines additionally overlap with the quadruplet of the ethoxycarbonyl group, which is possibly further split by a small long-range coupling to the phosphorus.

12.3.4 Carbon-13 NMR Spectrum

Besides the methyl and methylene carbon atoms in the ethoxycarbonyl group all carbon-13 signals are split into doublets due to ^{13}C–^{31}P couplings. The ^{13}C–^{31}P coupling constant over one bond is 133 Hz, those over two bonds are 6 and 7 Hz and the one over three bonds is 6 Hz. ^{1}H–^{1}H coupling constants generally decrease with increasing number of bonds between the coupling partners. In contrast, coupling constants between heavier nuclei often increase with increasing distance between the nuclei, pass over a maximum and then decrease again.

13 Problem 13

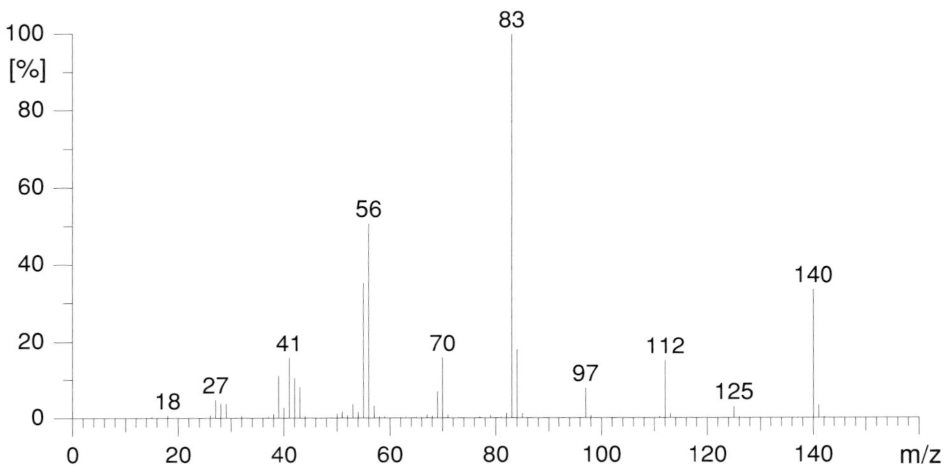

Fig. 13.1: Mass spectrum: EI, 70 eV

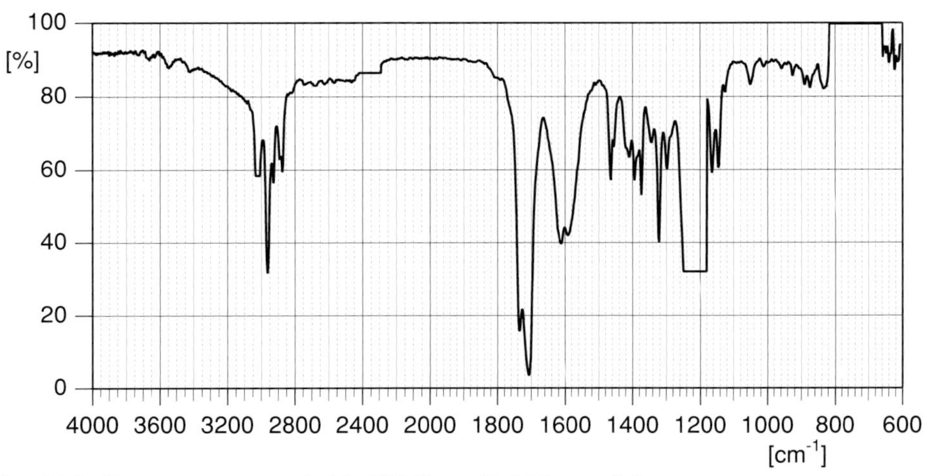

Fig. 13.2: IR spectrum: recorded in $CHCl_3$, cell thickness 0.2 mm

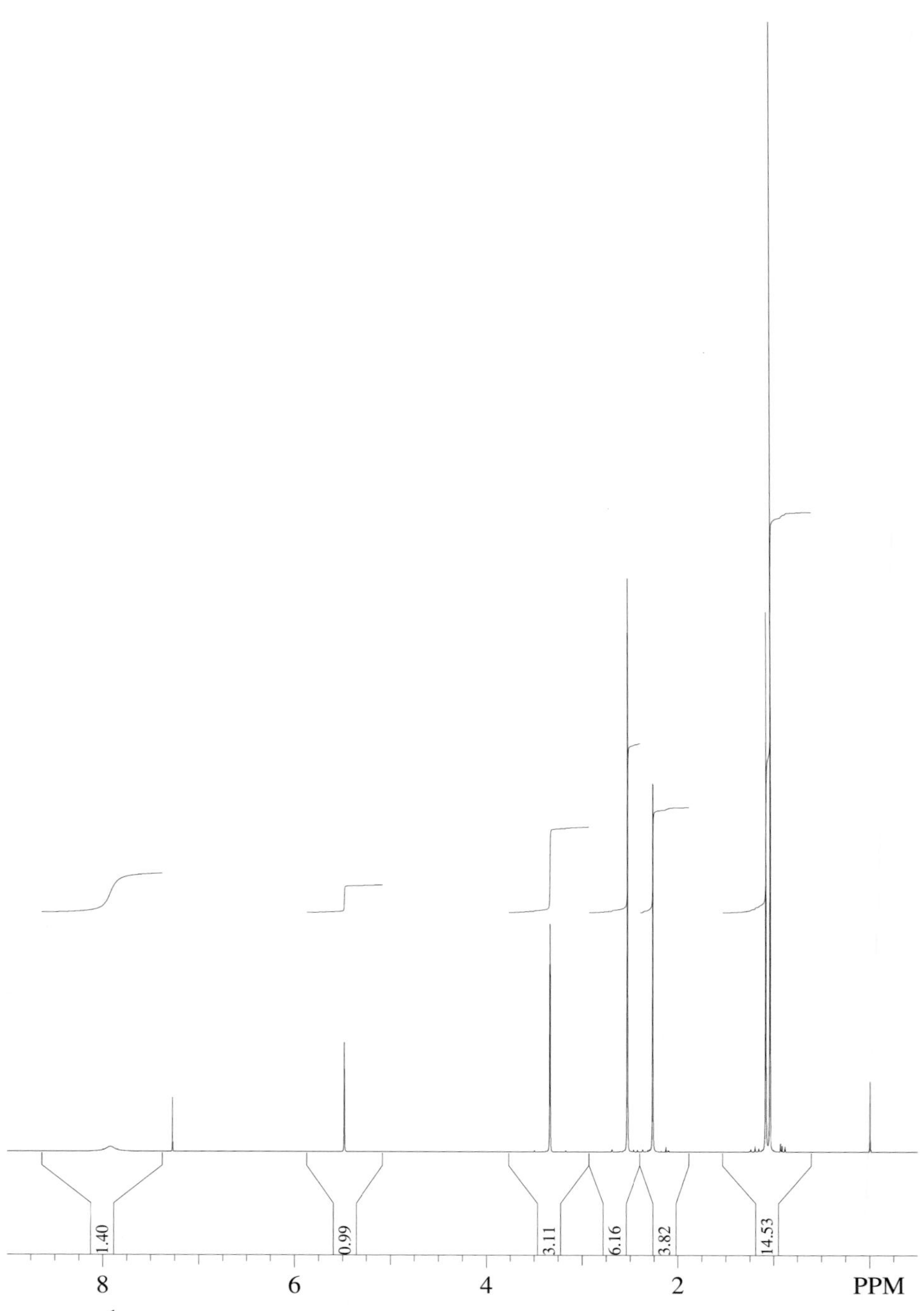

Fig. 13.3: ^{1}H-NMR spectrum: 400 MHz, solvent: $CDCl_3$

Fig. 13.4: ^{1}H-NMR spectrum: 400 MHz, solvent: CD_3OD

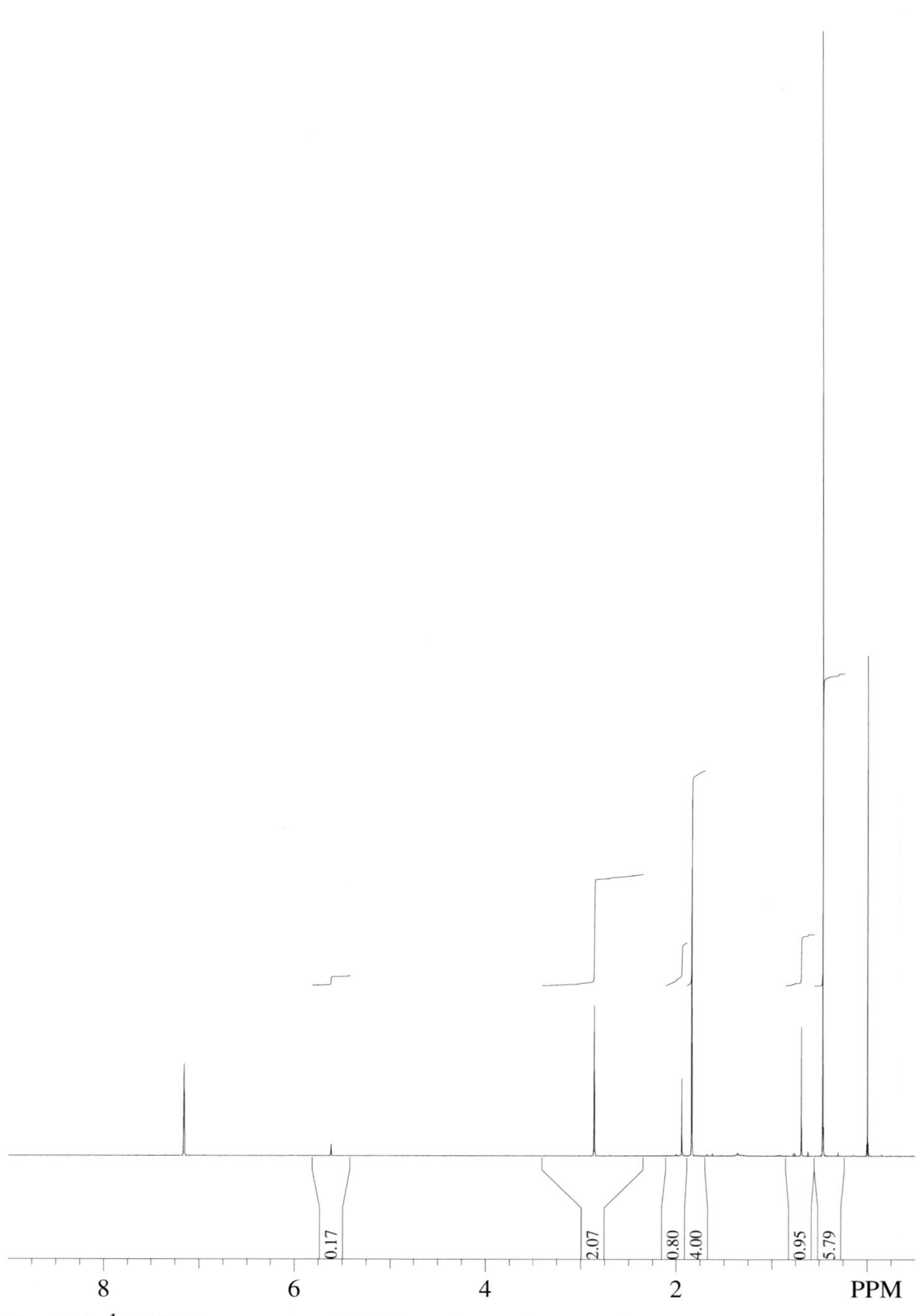

Fig. 13.5: ^{1}H-NMR spectrum: 400 MHz, solvent: benzene-d_6

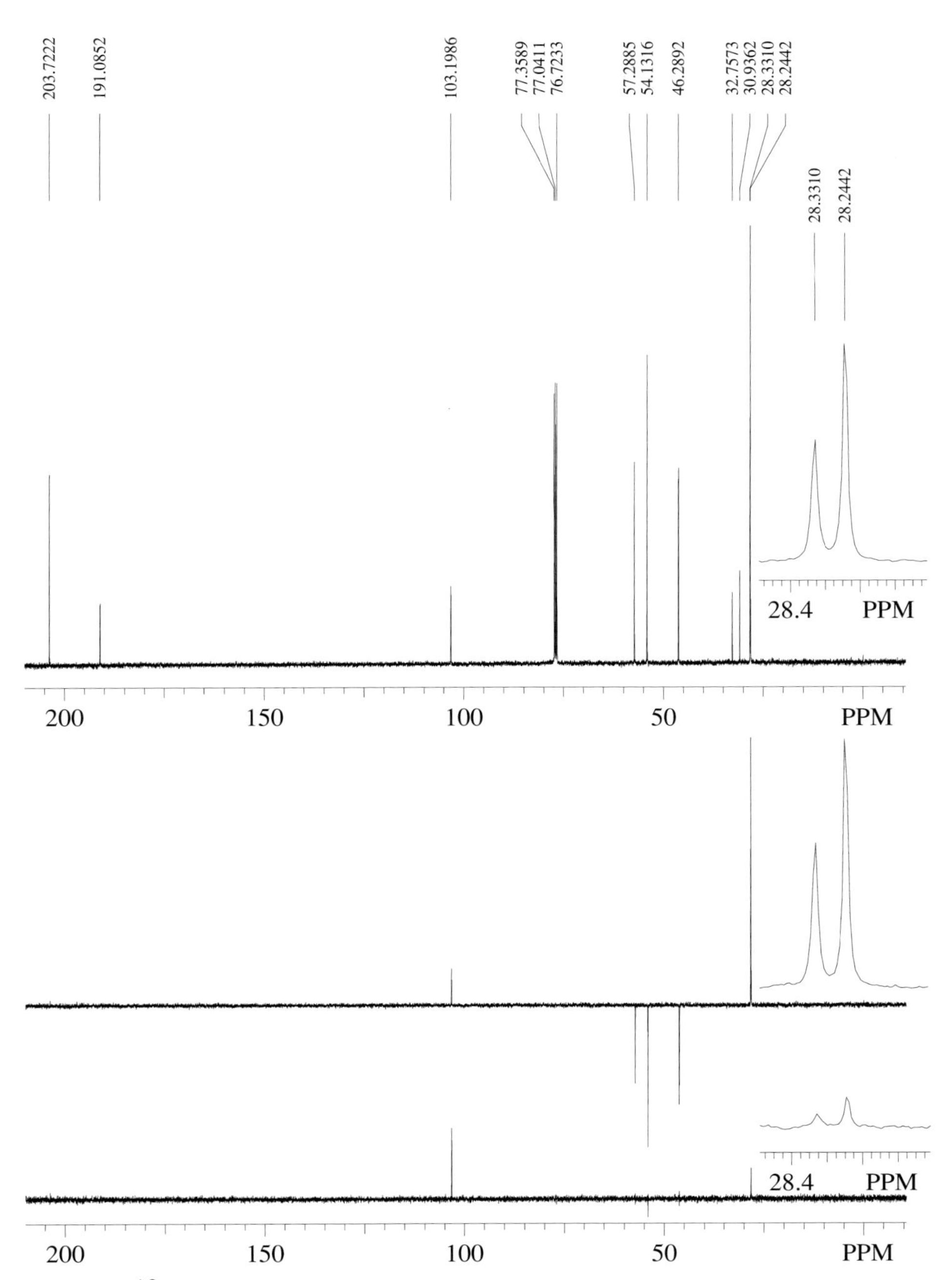

Fig. 13.6: ^{13}C-NMR spectrum: 100 MHz, solvent $CDCl_3$, Top: proton decoupled, middle: DEPT135, bottom DEPT90 ($\tau = 3.6$ ms)

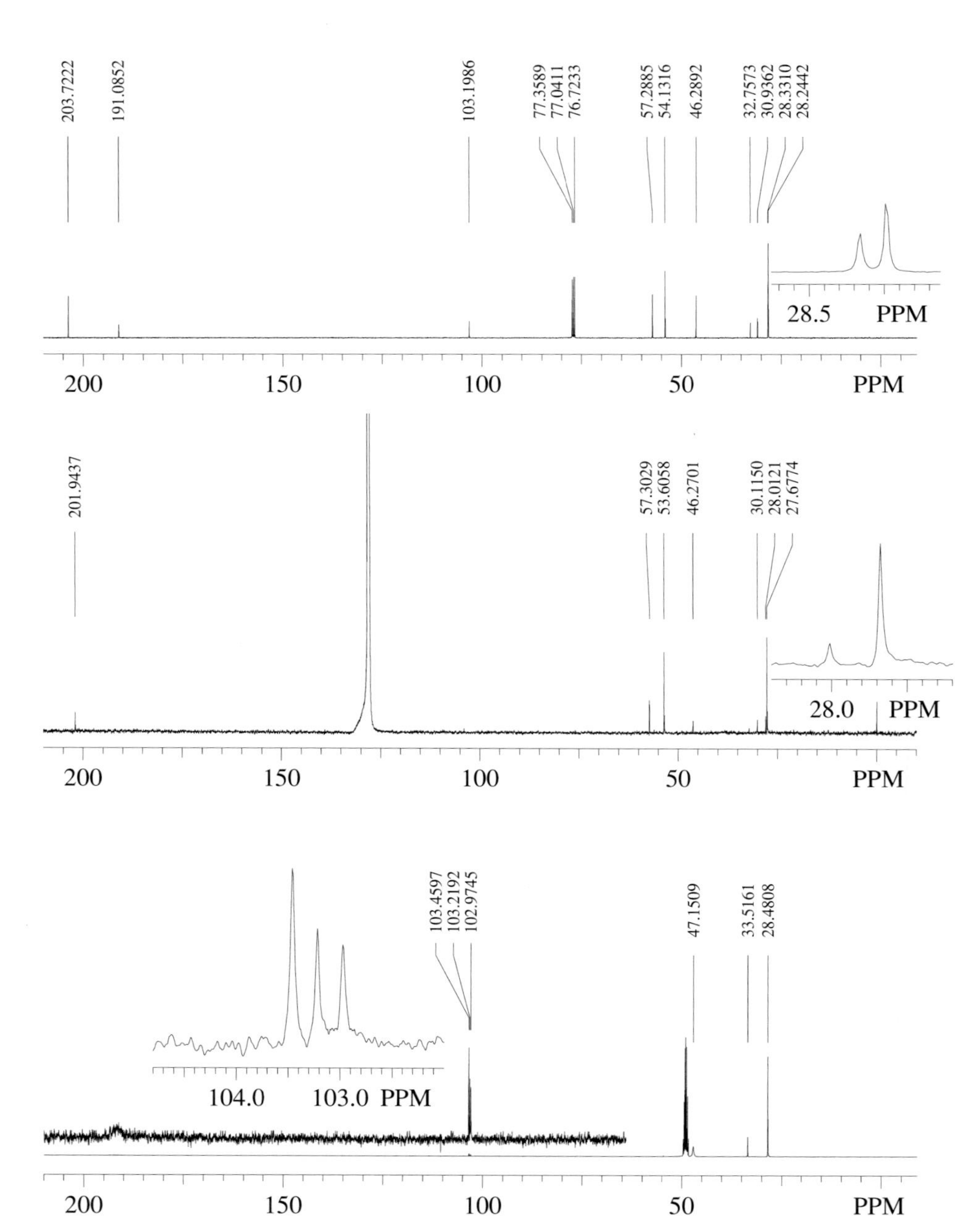

Fig. 13.7: ^{13}C-NMR spectra: 100 MHz, in various solvents (from top to bottom): $CDCl_3$, benzene-d_6, and CD_3OD

13.1 Elemental Composition and Structural Features

The mass spectrum ends with m/z 140 which is assumed to be the molecular mass. The first fragment peaks correspond to the loss of 15 (m/z 125), 28 (m/z 112) and 43 mass units respectively (28+15 leading to m/z 97). Intensity distribution and fragment masses indicate a nonaromatic unsaturated system without evidence for nitrogens.

In the infrared spectrum we find a structured broad absorption band in the range of 2300 to 3500 cm^{-1}. Bands of this type are generally observed for ammonium compounds (N^+–H stretching and combination bands) and for strongly hydrogen bonded O–H compounds (O–-H stretching vibrations) e.g., in carboxylic acids. We need at least two oxygen atoms in order to be able to form a strongly hydrogen bonded hydroxyl group. Indeed, we find carbonyl stretching vibration bands at 1705 and 1735 cm^{-1}. It is not possible to interpret the strong absorption bands at about 1600 cm^{-1} at this stage.

The intensity ratios in the proton NMR spectrum recorded in deuterochloroform as a solvent are rather confusing. We have either an impure sample or a poor quality of integration. The best approximation leads to the following ratios: 1 : 1 : 3 : 6 : 4 : 15, 30 protons in total. In deuteromethanol as solvent, we find three signals with an intensity ratio of 1 : 2 : 3. The low-field signal corresponds thereby to the OH proton of the solvent, which contains all exchangeable protons of the sample and, additionally, non-deuterated methanol and water from the solvent, if present. We notice that the chemical shifts of the two signals in deuteromethanol at δ = 1.1 and δ = 2.2 fit to two corresponding shifts in chloroform, both of which are split in the latter solvent. The signals at δ = 8.0, 5.5 and 3.4, on the other hand, do not appear in methanol. Since exchangeable protons are substituted by deuterium nearly quantitatively in this solvent (the molar concentration of the solvent being much higher than that of the sample), one could assume that the protons corresponding to these signals are exchangeable and are thus included in the signal at 5 ppm in methanol. The fact that the signals at 1.1 and 2.2 ppm are split in chloroform suggests that two different forms are present in the two solvents. The unusual intensity ratio of the split signals of ca. 5 : 8 would require a much too large number of protons which cannot be accommodated with the given molecular mass. Therefore, we assume the presence of two slowly exchanging tautomers or conformers in chloroform. The proton NMR spectrum recorded in deuterobenzene corroborates this interpretation by giving similar signals but strongly different relative intensities (ca. 1 : 6) indicating that we have basically the same equilibrium which is shifted to favor one form. In deuteromethanol only one of the two forms is present.

Comparing the signal intensities in the carbon-13 NMR spectra recorded in chloroform and in benzene allows us to assign the lines to the two forms. The one dominating in benzene with signals at 27.7, 30.2, 53.6, 57.3 and 201.9 ppm is responsible for the signals at 28.3, 30.9, 54.1, 57.3 and 203.7 ppm in chloroform. According to the DEPT spectra they correspond to CH_3, C, CH_2, CH_2, and C, respectively, and the integration of the corresponding signals in the proton NMR spectrum recorded in benzene gives the following proton counts (from left to right): 2 : 4 : 6.

13.2 Structural Assembly

The following structural fragments have thus been found for the form dominating in benzene:

Structural fragment	Mass
2 CH_3 (equivalent by symmetry)	30
2 CH_2 (equivalent by symmetry)	28
1 CH_2	14
1 C	12
1 C=O	28
1 O	16

These elements sum up to $C_7H_{12}O_2$ (mass 128). The difference to the molecular mass of 140 corresponds to one carbon atom. This must be equivalent by symmetry with one of the non-protonated carbon atoms found above, since no further signals in the carbon-13 NMR spectrum are available. The molecular formula $C_8H_{12}O_2$ indicates three double bond equivalents. The form in question does not contain any olefinic double bond. Thus we need two equivalent carbonyl groups and one ring to accomplish the required three unsaturations. The two carbonyl groups have to be arranged in such a way, that no vicinal couplings between the methylene protons occur. The only such possibility is dimedone:

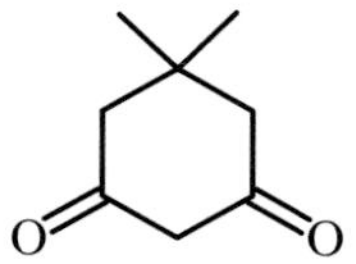

O O

The tautomerization of dimedone as a β-diketone explains that more than one form may occur with concentration ratios which are highly dependent on the solvent.

13.3 Comments

13.3.1 Mass Spectrum

Loss of 28 mass units from the molecular ion must be ascribed to a decarbonylation, probably involving a ring contraction reaction. If it were due to loss of ethylene, one would expect it to be accompanied by loss of ethyl radical (mass difference 29).

Formation of the base peak m/z 83 (and subsequent decarbonylation to m/z 55) can be visualised as arising by the typical cyclohexanone reaction sequence a → b → c → d.

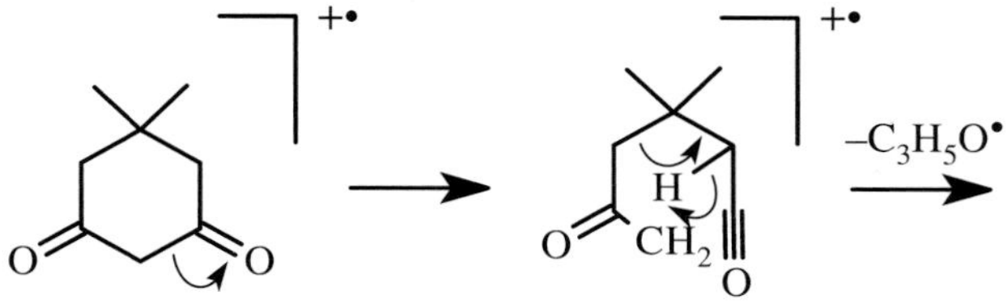
+•
O
O
+•
H
CH2
O
O
–C3H5O•

a b

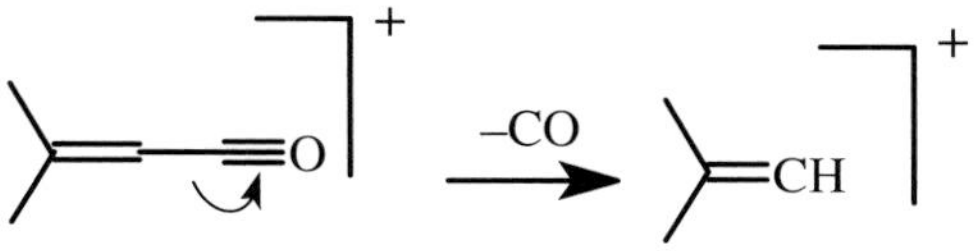
+
O
–CO
CH
+

m/z 83 m/z 55

c d

Formation of the prominent m/z 56 could be the result of a retro-Diels-Alder reaction in the enols **II A** and **II B** (see below) according to the following scheme:

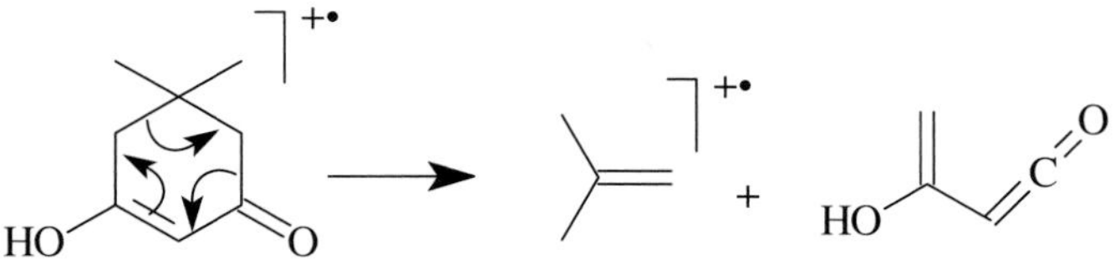
+•
HO
O
+•
+
HO
C
O

or else of elimination of two neutral ketene molecules, which are particularly good leaving groups in gas phase reactions, as follows:

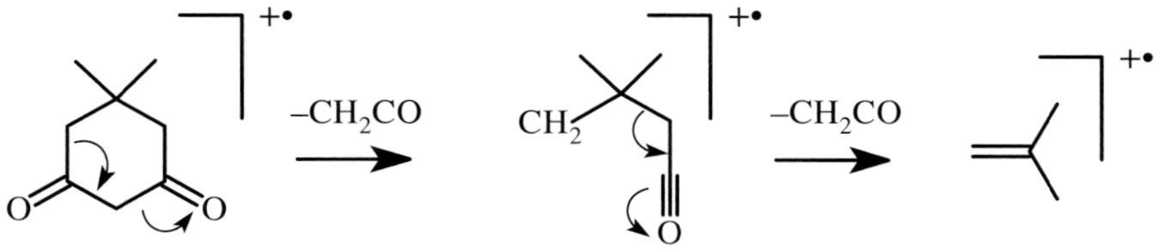
+•
O
O
–CH2CO
+•
CH2
O
–CH2CO
+•

m/z 56

13.3.2 Proton and Carbon-13 NMR Spectra

There are three possible tautomeric forms of dimedone having reasonable stability:

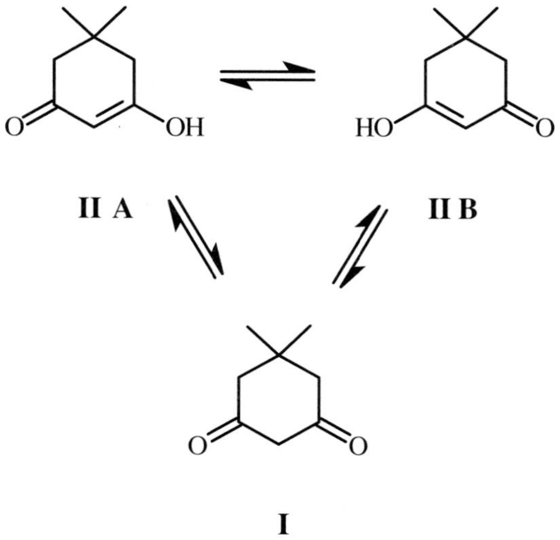

II A **II B**

I

In chloroform and benzene as solvents, the exchange rate between **I** and **II A** or **II B** is slow on the NMR time scales whereas the exchange rate between **II A** and **II B** is fast leading to magnetic equivalence of the corresponding nuclei caused by kinetic effects (Chapter 19.4). We would thus observe in solvents without exchangeable deuterium five carbon-13 signals and three proton signals for **I** and five carbon-13 and four proton signals for **II A / II B**. The methyl signals of **I** and **II A / II B** are almost equivalent in the carbon-13 NMR spectrum (solvent chloroform) and, therefore, only nine of the ten predicted signals are discernible in this spectrum. In chloroform, the chemical shift values of $\delta = 191.1$ correspond to the mean value of the shifts for =C–O and C=O in **II A / II B**. The signal at $\delta = 103.2$ can be assigned to the second alkene carbon atom of these tautomers. In benzene, the concentration of this tautomers is so small that some of the signals are not seen in the spectra.

In deuteromethanol as solvent, all protons involved in the tautomeric reactions are exchanged with deuterium. These are the alkene and the hydroxyl proton of **II A / II B**. The population of the keto form **I** is very low in this solvent, with the result that no corresponding signals are detectable in the NMR spectra presented here. The carbon-13 NMR spectrum is complicated by two different facts in this solvent. First, the alkene methine is now predominantly deuterated. Deuterium having a spin quantum number of I = 1 leads to splitting of the signal of the coupling partners into 3 lines with relative intensities of 1 : 1 : 1. This coupling is, of course, not influenced by proton broad band decoupling. The coupling constant is smaller than the corresponding C–H coupling constant by a factor of 6.514 as demanded by the gyromagnetic constants of proton *versus* deuterium. The spectrum shows a significantly higher intensity of the low-field line of the triplet (at 103.5 ppm). This can be explained by the presence of a small amount of non-deuterated sample. Its CH signal is at lower field due to the isotope effect of deuterium. Since the relaxation times of the non-deuterated and deuterated CH are expected to be significantly different, it is not possible to directly interpret the signal intensities.

Another surprising feature of the carbon-13 NMR spectrum in methanol is the low intensity as a consequence of the large line-width of the signal at 190 ppm. This can be explained by the limited exchange rate between **II A** and **II B** so that the =C–O and C=O signals are not fully averaged. The same effect operates on the CH_2-signals at 47.2 ppm. However, the line width is considerably smaller in this case. This indicates that the chemical shift difference between the two methylene carbons in **II A / II B** is smaller than that between the =C–O and C=O, as would be expected.

Check it in SpecTool:
Use the chemical shift estimation tools located under **TOP**, **HNMR Tools**, **^{1}H-Shift estimation** and **TOP**, **CNMR Tools**, **^{13}C-Shift estimation** to estimate the chemical shifts in **I** and **II**, and verify the assignments. The same structure input can be used for ^{1}H- and ^{13}C-NMR prediction. Note that the estimation of the chemical shifts of quaternary carbon atoms is often unreliable.

13.3.3 Infrared Spectrum

Infrared spectral data corroborate the facts deduced from the NMR spectra, namely that in chloroform a mixture of keto- and enol form is present. In the keto form the stretching vibrations of the two carbonyl groups are coupled, leading to a double band at 1730 and 1705 cm^{-1}. The first, less intense band corresponds to the symmetric stretching mode, whereas the second strong band is due to the asymmetric vibration. In the enol, it is not possible to separate the various CC and CO vibrations, the strong bands at 1600 cm^{-1} are rather to be ascribed to the conjugated chromophore as a whole. In 2-monosubstituted or nonsubstituted 1,3-cyclohexadiones one often observes two bands for the enol. One is around 1640 cm^{-1} and corresponds to the free enol. The other band at about 1600 cm^{-1} is assigned to a dimer of the following type:

O···H—O

O—H···O

In chloroform, the enolized dimedone exists predominantly as a dimer, the monomer band at 1630 cm^{-1} is not discernible. Furthermore, if appreciable amounts of the 'free' enol were present, we would expect a moderately sharp band for the O–H stretching vibration above 3000 cm^{-1}.

14 Problem 14

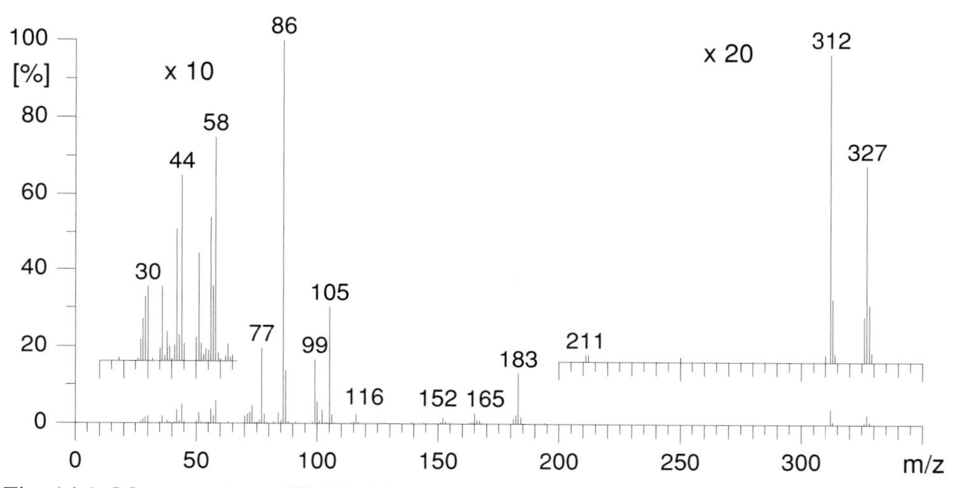

Fig. 14.1: Mass spectrum: EI, 70 eV

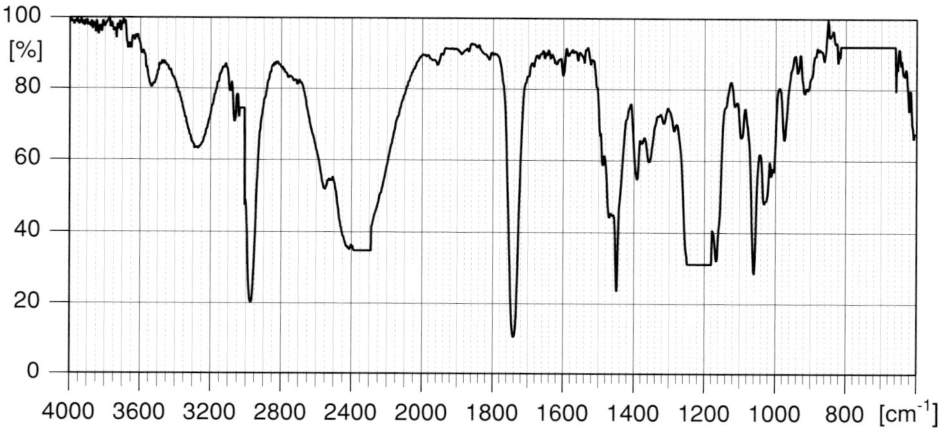

Fig. 14.2: IR spectrum: recorded in $CHCl_3$, cell thickness 0.2 mm

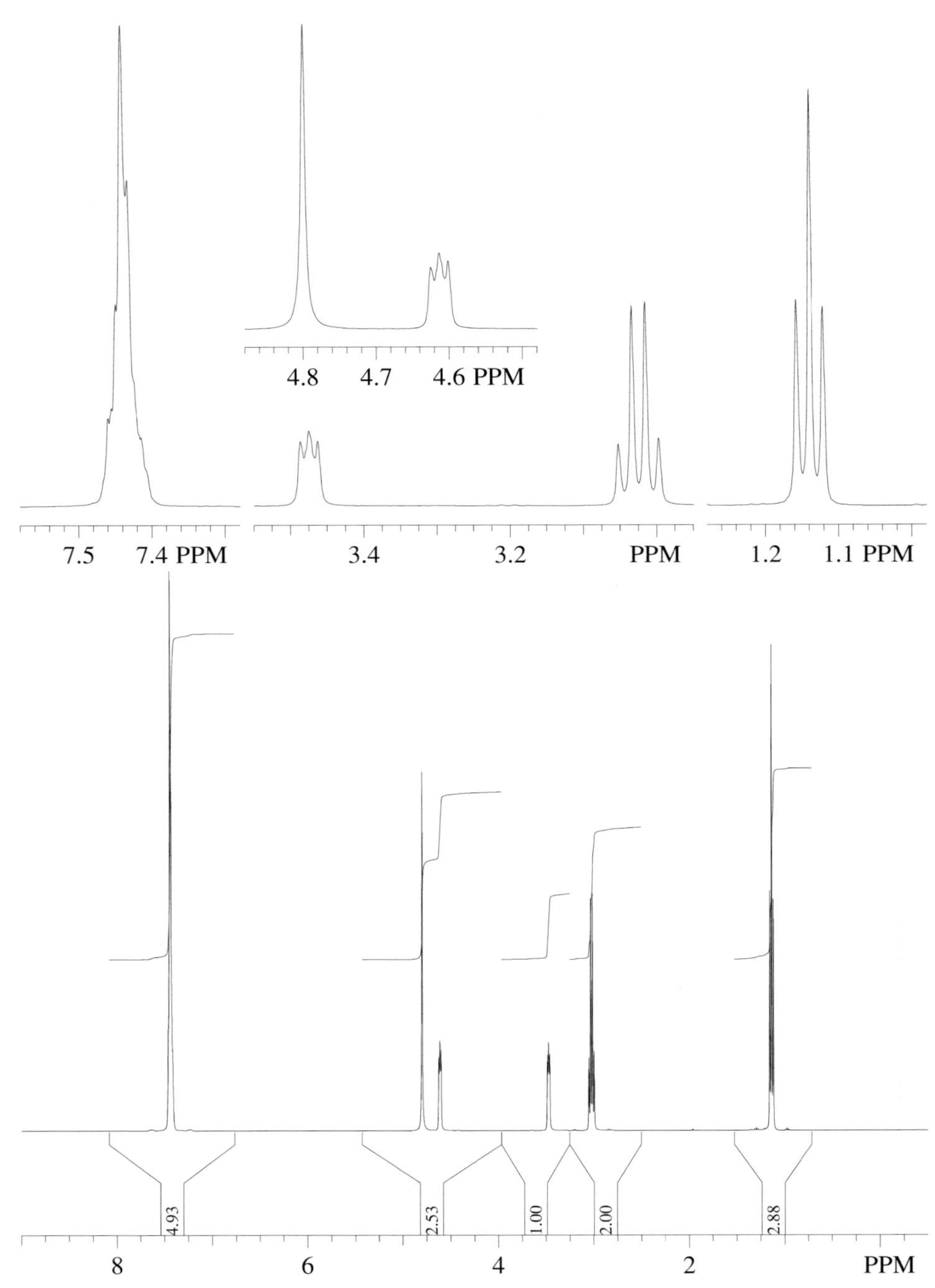

Fig. 14.3: ^{1}H-NMR spectrum: 400 MHz, solvent: D_2O

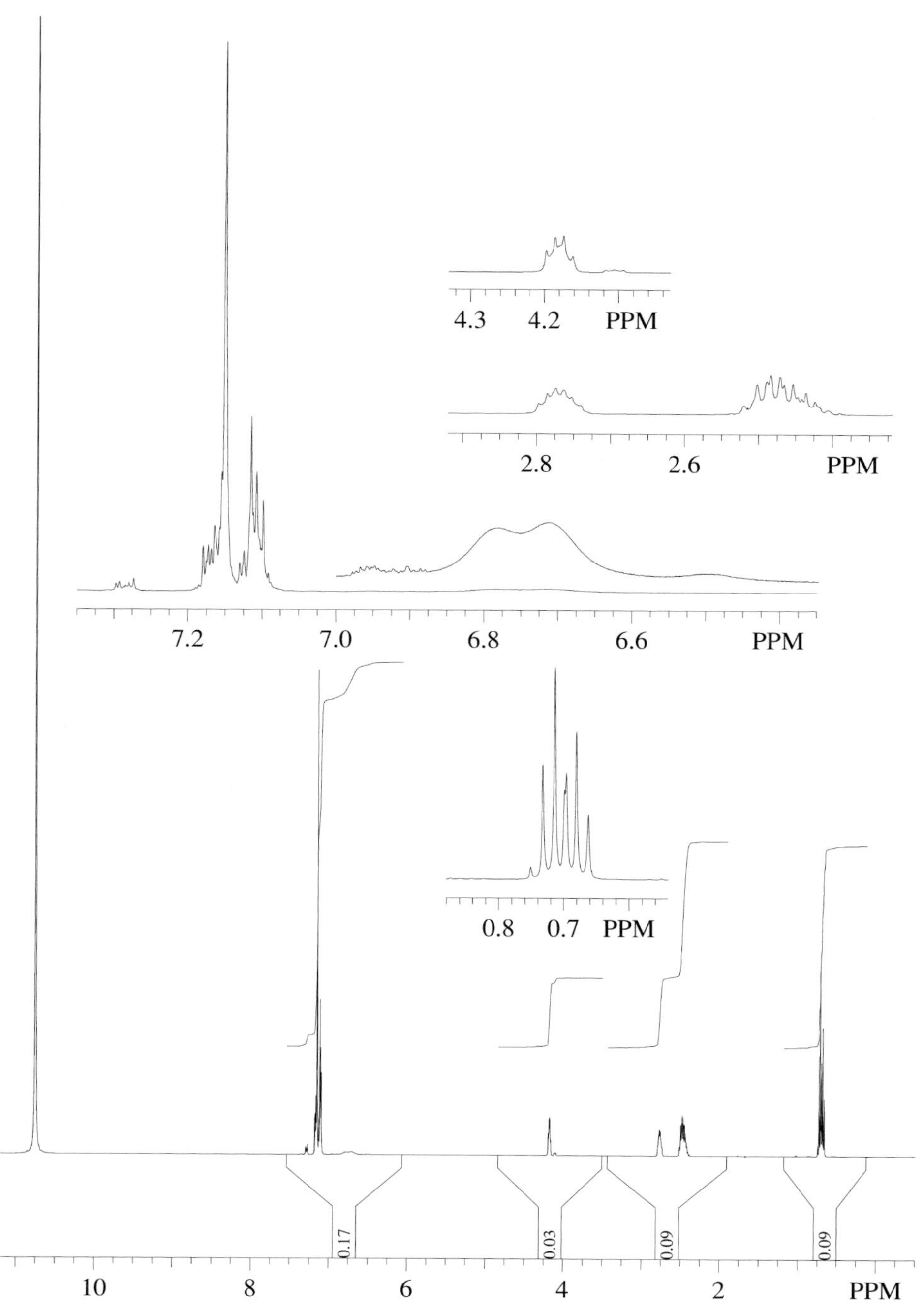

Fig. 14.4: ^{1}H-NMR spectrum: 400 MHz, solvent: benzene-d_6 and trifluoroacetic acid

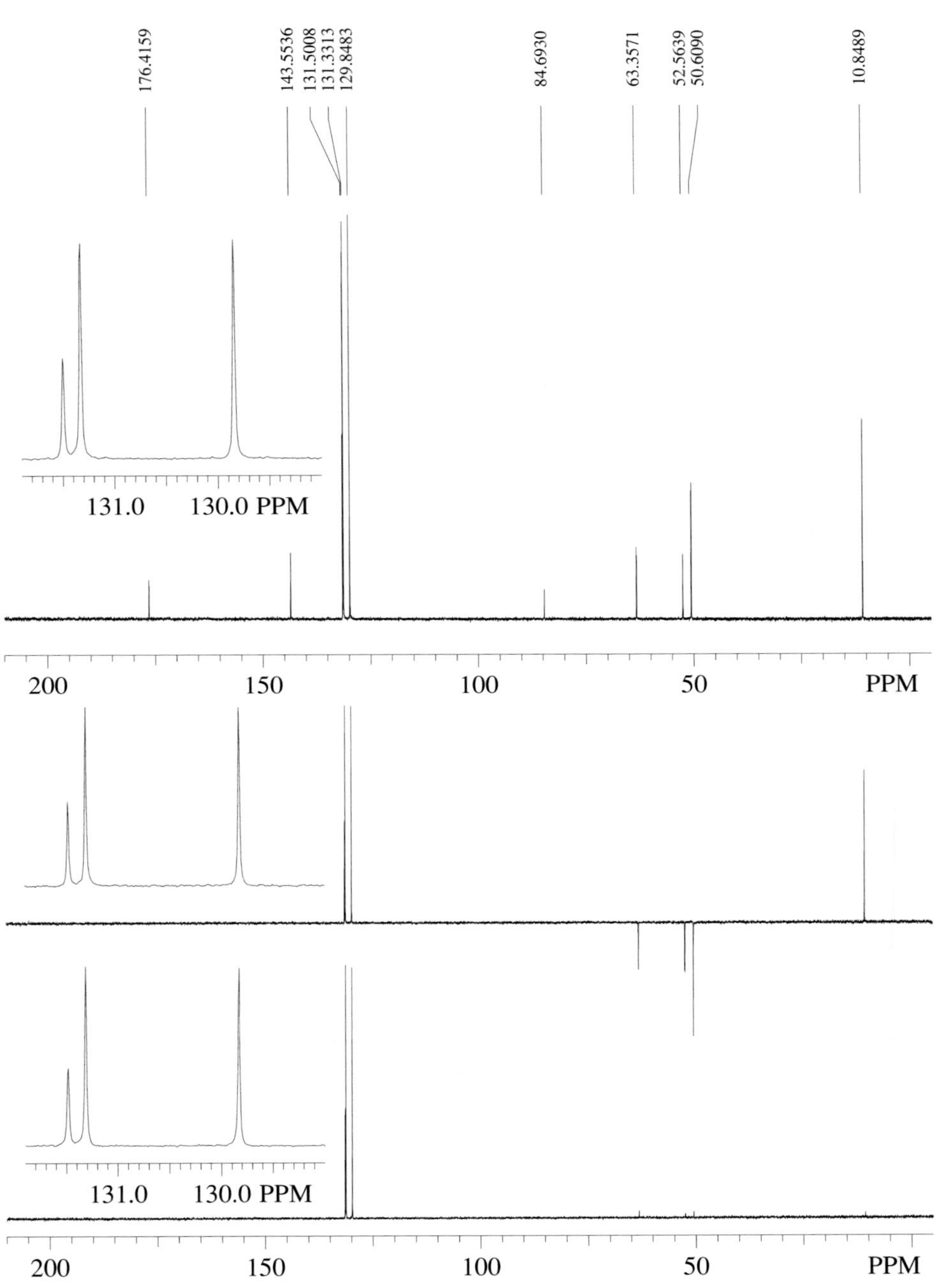

Fig. 14.5: ^{13}C-NMR spectrum: 100 MHz, solvent D_2O, Top: proton decoupled, middle: DEPT135, bottom DEPT90 ($\tau = 3.6$ ms)

14.1 Elemental Composition and Structural Features

Mass spectrum: m/z 327 appears to represent the molecular ion, since all peaks at lower mass show chemically reasonable mass differences relative to this signal. The odd mass value indicates the presence of an odd number of nitrogen atoms, the presence of nitrogen being corroborated by a small but significant fragment at m/z 30. Signals at m/z 36 and m/z 38 of proper intensity ratio show the presence of HCl. Since no chlorine atom is evident in what is assumed to be the molecular ion (no signal of about 30% relative intensity at m/z 329) the possibility of dealing with an ammonium salt of an amine with hydrochloric acid must be considered, unless one assumes HCl to be an impurity. The fragment series m/z 51, 77, 105, 183 (105 + 78) and m/z 30, 44, 58, 86, respectively, reveal a mixed aromatic/nonaromatic nature of the sample.

> **Check it in SpecTool:**
> Use **MS Ranges** and **MS Tools** to verify the assumptions and assignments made. Navigate to the mass spectrum of this problem in the **SpecLib** section amplify the region of small signals below m/z 77.

Infrared spectrum: Two bands at 3540 and 3270 cm^{-1} indicate a hydroxyl group (OH stretching vibrations of free and hydrogen bonded hydroxyl group). An NH stretching vibration at 3540 cm^{-1} is very unlikely and would give rise to a sharp absorption band. The broad band at 2400 cm^{-1} is taken as evidence corroborating the presence of an ammonium salt (see Comment), the band at 1740 cm^{-1} shows a carbonyl group.

> **Check it in SpecTool:**
> The band at 2400 cm^{-1} has a peculiar shape. Could this be due to solvent absorption? Check it in SpecTool by using e.g., a spectrum of an ammonium salt in the **IR Spectra** section.

Proton NMR spectrum: The integral ratios are: 5 (singlet), ~1.5 (singlet), 1 (multiplet), 1 (multiplet), 2 (quartet), 3 (triplet). The signals at 1.1 and 3.0 ppm clearly correspond to an ethyl group without any coupling to other parts of the molecule. It must be attached to the nitrogen to explain the chemical shift of the methylene group. Since the multiplets at $\delta = 4.6$ and $\delta = 3.5$ apparently are caused by coupling of the respective protons with one another, they cannot be due to one proton each. The numbers in the integral ratios must, therefore, be doubled. The singlet of ten hydrogen atoms at $\delta = 7.4$ by its sharpness suggests two equivalent phenyl groups, which should be attached to an sp^3 hybridized carbon atom, because all hydrogen atoms are nearly isochronous. The signal at 4.8 includes the exchangeable protons which must be present according to infrared evidence and HDO introduced as impurity with the solvent D_2O. Thus its integral is probably too high. The multiplets at $\delta = 4.6$ and $\delta = 3.5$ by their symmetrical arrangement and number of hydrogen atoms constitute an *AA'BB'* spin system and must thereby be attributed to a pair of methylene groups. Chemical shift values require that both methylene groups be attached to deshielding centers.

Carbon-13 NMR spectrum: Only ten nonisochronous carbon atoms are discernible in the spectrum. Some of these must be doubled if the conclusion drawn from the proton NMR spectrum is correct, implying that equivalent functional groups are present. In addition to those carbon atoms which must be expected from the already identified functional groups one can recognize a quaternary carbon atom (cf. DEPT spectrum) attached to a hetero atom (δ = 84.7). The carbonyl carbon atom is also shown by its chemical shift (δ = 176.4) to be bonded to a hetero atom.

Check it in SpecTool:
The chemical shifts of various carbonyl functionalities can be found under **CNMR Data, Special, C=O**. Verify that the chemical shift is not valid for a keto carbonyl group. A quick way to do this is to click at the various table headings and noting whether the distribution diagram at the bottom shows lines in the respective region.

The information thus acquired shows the following structural elements and related masses:

2 (phenyl)	154	–N–H	15
C (quaternary)	12	–O–H	17
$-CH_2-CH_2-$	28	–C=O	28
2 $-CH_2CH_3$	58	–Cl	35

The mass of these structural elements adds up to 347 mass units. If the sample is in fact an ammonium chloride and m/z 327 represents the molecular ion of the corresponding base, then the complete molecular mass is 327 + 36 (HCl) = 363 and the deficiency of the mass balance above makes 16 mass units, which must be attributed to one oxygen atom. The molecular formula is then $C_{20}H_{25}NO_3 \cdot HCl$ and shows nine double bond equivalents, which are taken care of by two phenyl groups and one carbonyl group.

14.2 Structural Assembly

The two ethyl groups must be attached to the nitrogen atom, because the chemical shift of their methylenes requires a hetero atom next to them and because they can only become equivalent if they are attached to the same hetero atom. Nitrogen, in turn, must be attached to the pair of methylenes to explain the chemical shift of a third methylene group and the appearance of the dominant even mass base peak at m/z 86 in the mass spectrum. The phenyl groups are terminal functions and can only be hooked to the quaternary carbon atom, because equivalence of the groups and connection to sp^3

hybridized carbon is required. The quaternary carbon atom must carry in addition the hydroxyl group in order to make the mass of the aromatic moiety 183 and explain the aromatic fragment series m/z 183, 105, 77, 51. The elimination of benzene (183 $\rightarrow$ 105) requires a mobile hydrogen atom and the following transition m/z 105 $\rightarrow$ 77 must be due to a decarbonylation to yield a phenyl cation.

To complete the molecule the diphenylcarbinyl and the diethylamine part must be connected via CO_2 resulting in two possible constitutions **I** and **II**:

$(CH_3CH_2)_2NH^+–CH_2CH_2–O–C(=O)–C(OH)(C_6H_5)_2 \; Cl^-$

I

$(CH_3CH_2)_2NH^+–CH_2CH_2–C(=O)–O–C(OH)(C_6H_5)_2$

II

A decision in favor of **I** can easily be made by estimating the expected proton and carbon-13 chemical shifts.

Check it in SpecTool:
Access **^{1}H-Shift** from the **HNMR Tools** page and estimate the chemical shifts.

Other arguments in favor of **I** are the fragments at m/z 116 and m/z 98 in the mass spectrum (see Comments).

14.3 Comments

14.3.1 Mass Spectrum

Ammonium salts are not significantly volatile. Those of amines with strong acids usually yield upon heating the spectrum of the free base along with that of the free acid, the latter frequently appearing at low intensity. HCl in this case is easily recognized only because the mass range m/z 35 to 37 is usually empty. Alkyl ammonium halides can give products of Hoffmann degradations and analogous reactions while complex anions are usually difficult or impossible to identify. As nitrogen is an excellent carrier of the positive charge, the spectrum of aliphatic amines is usually dominated by the largest nitrogen-containing fragment that can arise by α-cleavage, here m/z 86:

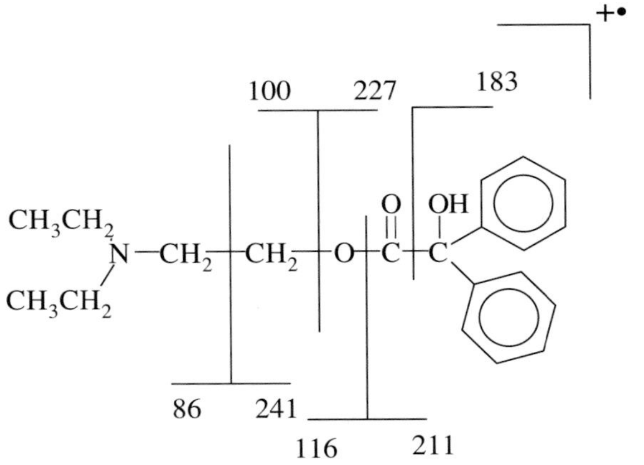

Check it in SpecTool:
Look for a suitable reference compound in the **MS Spectra** section to rationalize the fragments of the amine part of the compound.

The signal at m/z 99 can be explained by a McLafferty rearrangement involving the ester carbonyl group, where the nitrogen containing part grasps the charge.

Check it in SpecTool:
A scheme of the McLafferty rearrangement for esters can be found in **MS Data**, **COO**, **Ester Fragmentation Scheme**.

14.3.2 Infrared Spectrum

Broad bands in the range 2000 to 3000 cm^{-1} are indicative for N-H stretching and combination vibrations of ammonium salts, with the main maximum appearing below 2500 cm^{-1} only if tertiary amines are involved.

14.3.3 Proton NMR Spectrum

The chemical shifts are defined relative to tetramethylsilane (TMS) which is dissolved in the sample. However, another procedure is needed with water as solvent since TMS is not sufficiently soluble in it. Usually, either water soluble derivatives are applied or TMS is added as external reference in a thin capillary. In this case the local magnetic fields in the capillary and in the sample are different due to the different volume susceptibilities of the two media. For cylindrical capillaries, a mathematical correction can be used in good approximation by using the volume susceptibilities of the pure solvent and TMS. Another method was utilized in the present case: The spectrometer was calibrated with a different sample which only contained water and a water-soluble TMS derivative, and the settings were then applied to the actual measurement. Note, that such methods only yield approximated chemical shifts.

Check it in SpecTool:
Volume susceptibility correction terms can be calculated by using the corresponding **HNMR Tool**. Check that they are not the same for electro- or permanent magnets and superconducting magnets, the direction of the magnetic field relative to the probe being different.

The chemical shift of ammonium NH-protons lies generally in the range $\delta = 6$ to $\delta = 8$. The signals are often broad because of the $^{14}N-^{1}H$ coupling which is partially decoupled through the quadrupole relaxation of the nitrogen nuclear spin. The exchange rate between ammonium NH protons and other exchangeable protons is generally slow on the NMR time scale. If deuterated solvents with exchangeable deuterium atoms are applied (like heavy water in the present problem), deuterium is in a first approximation statistically distributed over the different exchangeable positions. Since the molar concentration of the solvent is very much higher than that of the sample, the exchangeable protons of the sample are nearly quantitatively substituted by deuterium, and the protons are located on the solvent molecules (in this case predominantly as HDO). The integrated intensity is generally higher than it would be for the protons of the sample, since such solvents are not fully deuterated.

In case of a slow exchange of the ammonium hydrogen/deuteron the nitrogen inversion is also slow. Then, the geminal CH_2 protons of the ethyl group are not equivalent and an ABX_3 spin system is expected. Due to symmetry, the spin systems are identical in the two ethyl groups. In the actual spectrum an A_2X_3 system is observed so that either the exchange (and nitrogen inversion) is fast, or the chemical shift difference between the two methylene protons too small to be detected. The exchange rate can be reduced by recording the spectrum in an acid such as trifluoroacetic acid as solvent. The present compound, however, spontaneously reacts with the solvent by forming an ester. Therefore, the spectrum shown in Fig. 14.4 corresponds to a mixture.

15 Problem 15

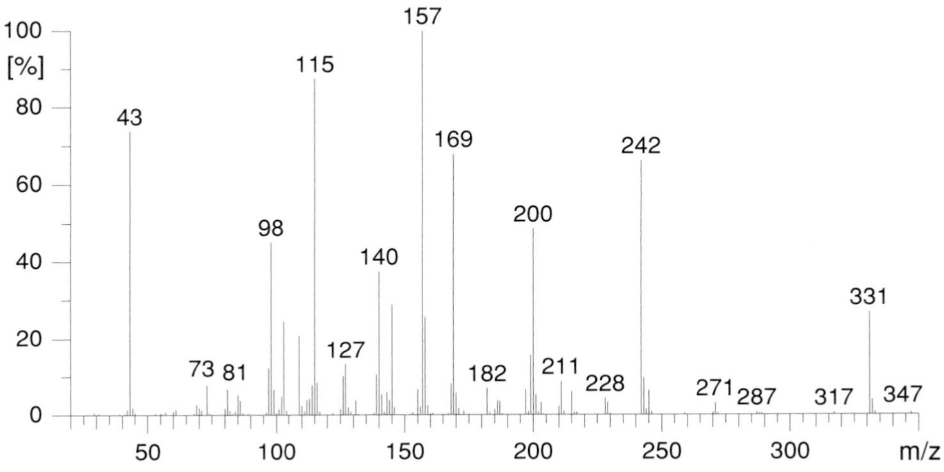

Fig. 15.1: Mass spectrum: EI, 70 eV

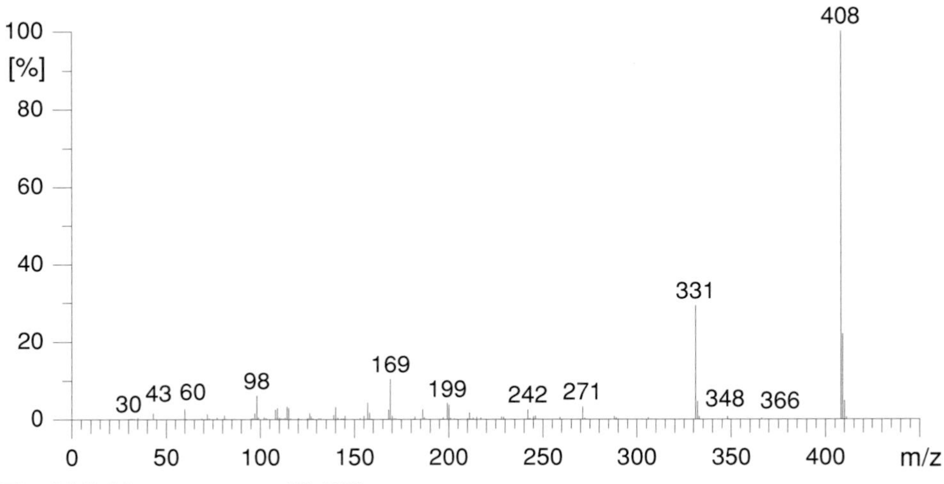

Fig. 15.2: Mass spectrum: CI, NH_4

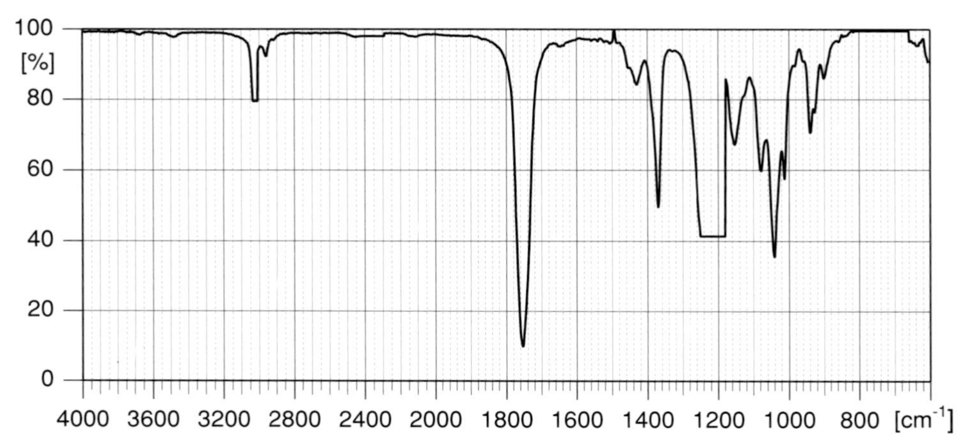

Fig. 15.3: IR spectrum: recorded in $CHCl_3$, cell thickness 0.2 mm

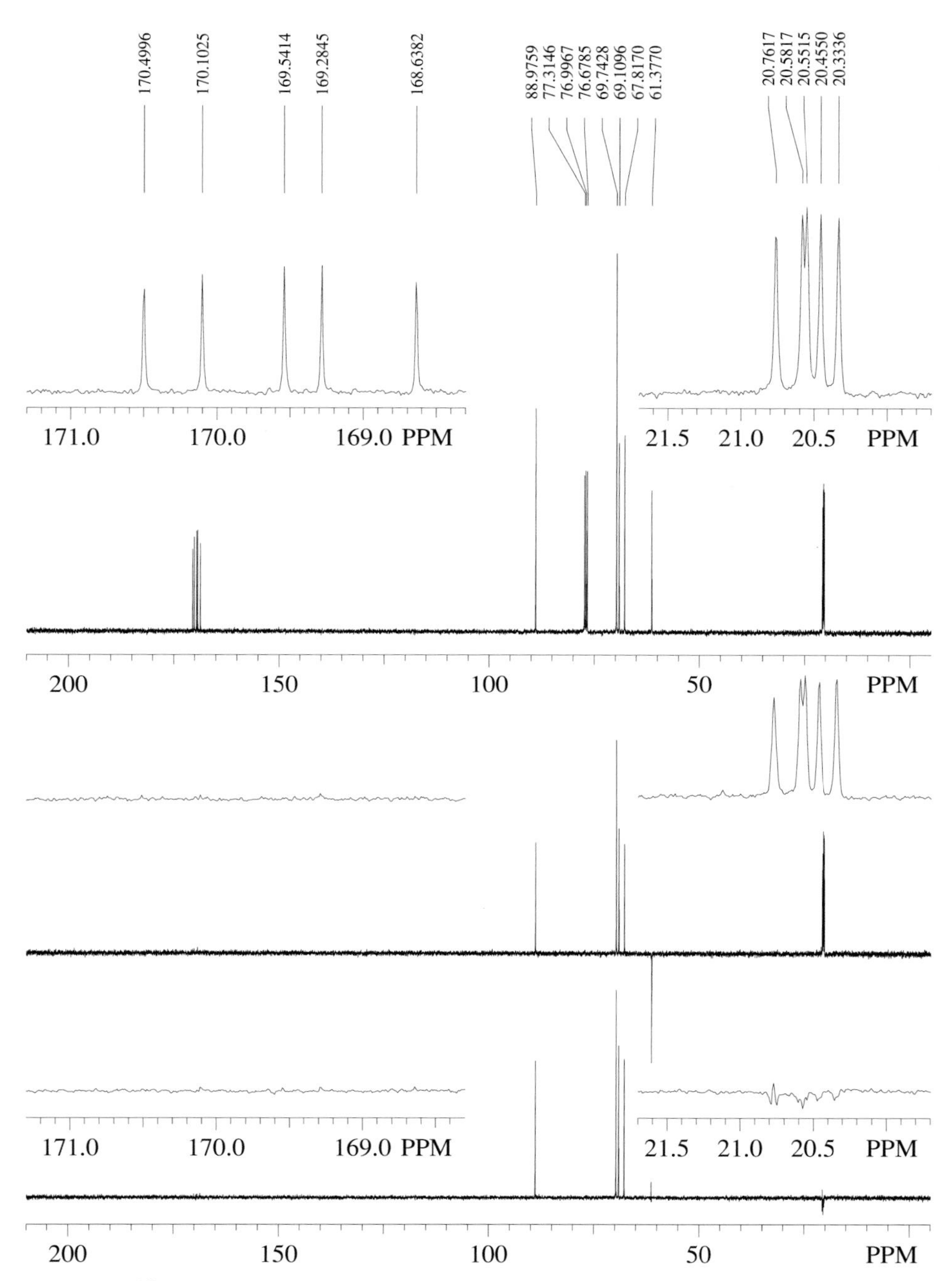

Fig. 15.4: ^{13}C-NMR spectrum: 100 MHz, solvent $CDCl_3$, Top: proton decoupled, middle: DEPT135, bottom DEPT90 ($\tau = 3.6$ ms)

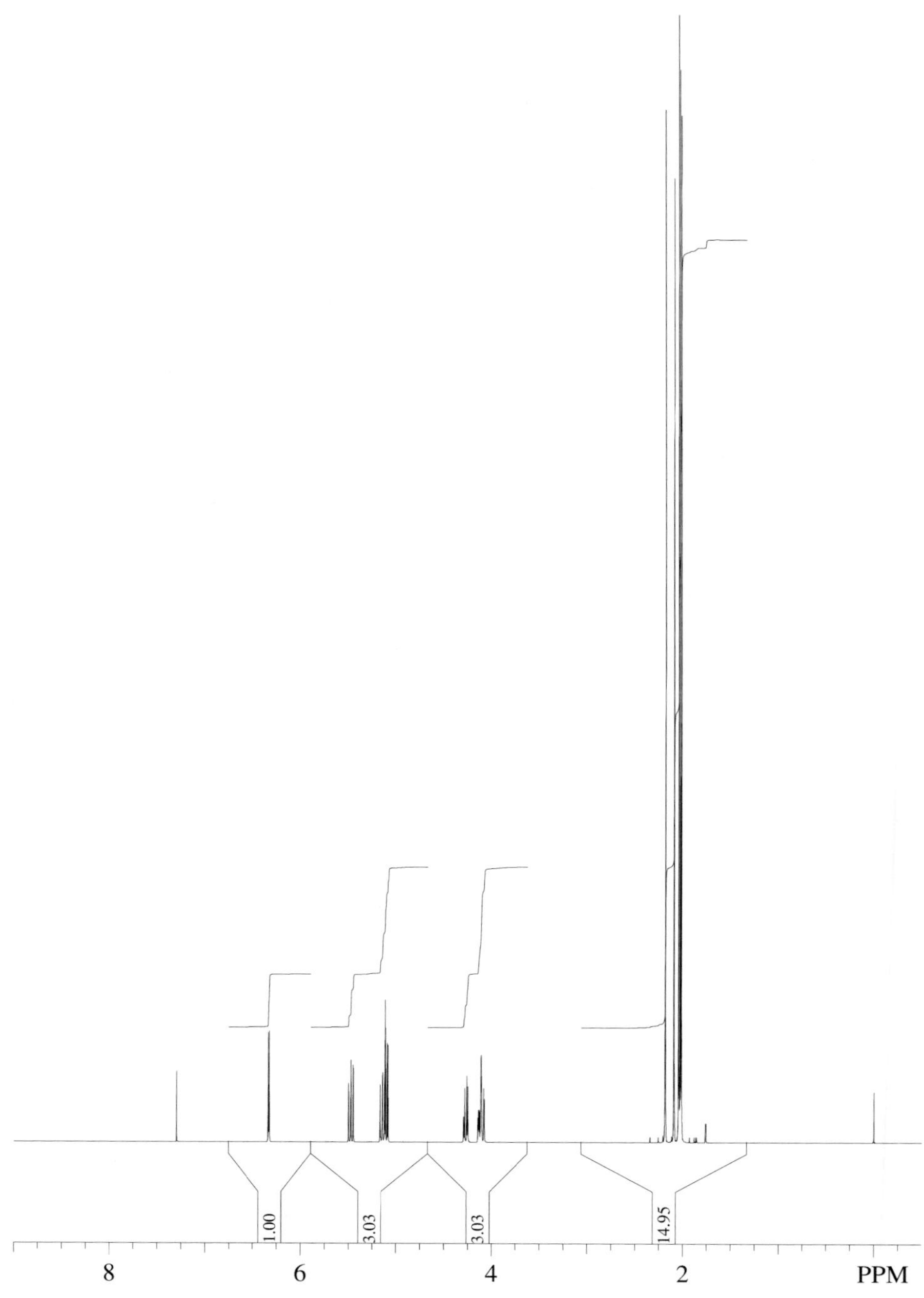

Fig. 15.5: ^{1}H-NMR spectrum: 400 MHz, solvent: $CDCl_3$

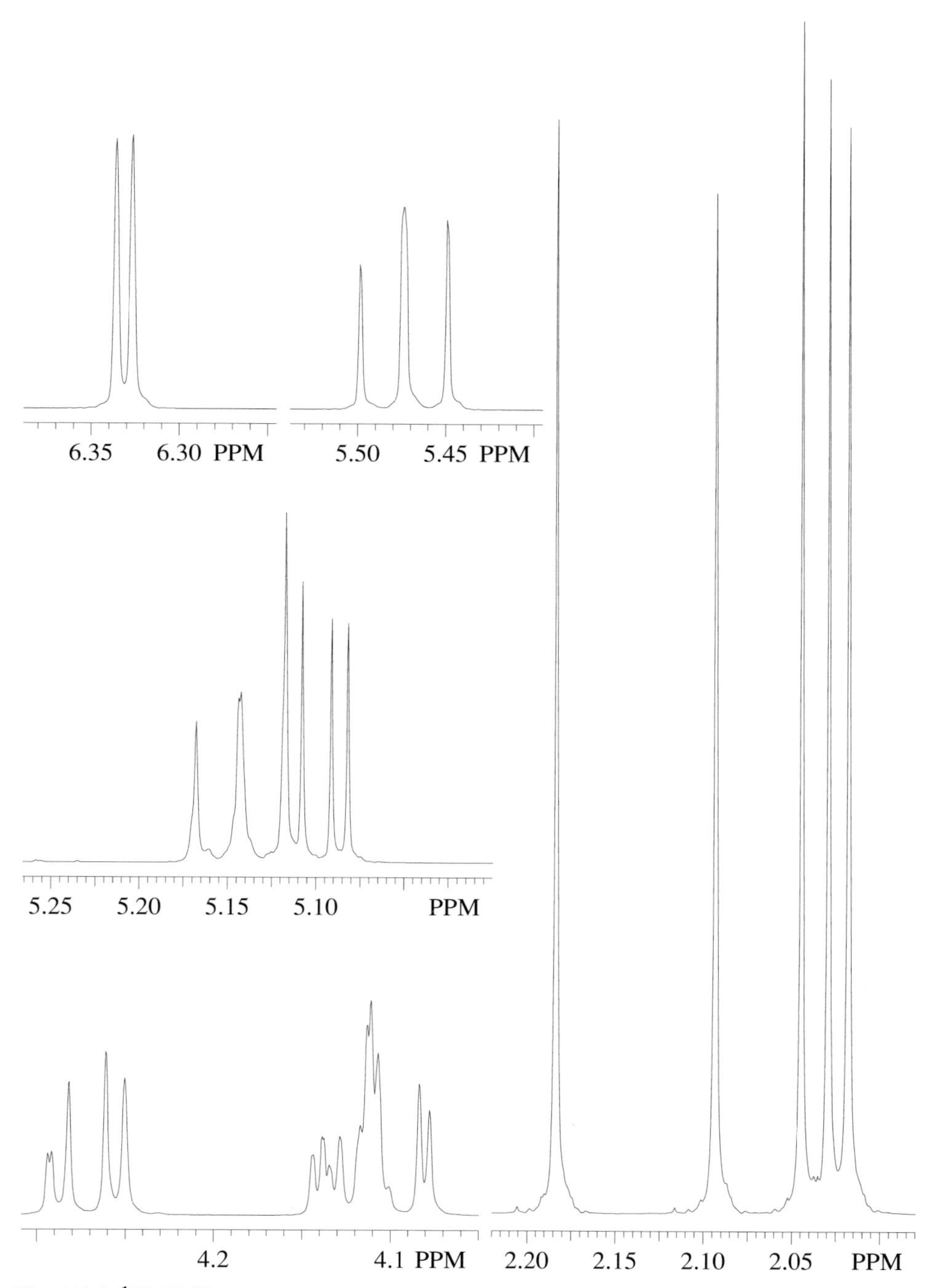

Fig. 15.6: ^{1}H-NMR spectrum (expanded regions): 400 MHz, solvent: $CDCl_3$

15.1 Elemental Composition and Structural Features

In the infrared spectrum the carbonyl stretching vibration band at 1755 cm^{-1} together with two prominent bands in the region of 1000 to 1300 cm^{-1} (recognizable despite the solvent absorption, possibly C–O–C stretching vibrations) suggest that the compound is an ester. There are no other characteristic absorptions which provide significant information at this stage.

The last important signal in the mass spectrum is at m/z 347. This peak is probably not the molecular ion since the mass difference to the next fragment (m/z 331) is 16 mass units (see Comments). If the molecular ion is not evident, a chemical ionization (CI) spectrum will help. In the present case the CI spectrum with ammonia as reactant gas indicates m/z 408 for $[M+NH_4]^+$ giving a molecular mass of 390. The fragments at m/z 347 and 331 correspond thus to $[M\text{-}43]^+$ and $[M\text{-}59]^+$. This is in line with the rather intense peak at m/z 43 without significant amounts of m/z 41 which suggests the presence of an acetyl group in the molecule.

The integration of the proton NMR spectrum results in the following intensity ratios: 1 : 1 : 2 : 1 : 2 : 3 : 3 : 9 and corresponds to a total of 22 hydrogen atoms in the molecule. In the narrow range of $\delta = 2.0$ to $\delta = 2.2$ there are five signals corresponding to 15 protons altogether. The chemical shifts fit well for acetyl methyl groups as already inferred from the mass spectrum. As all five methyl groups have very similar chemical shift values we suspect the presence of five acetyl groups.

In the carbon-13 NMR spectrum, we indeed find five carbonyl signals of very similar chemical shift values ($\delta = 168.6$ to $\delta = 170.5$). These values exclude ketones and are in very good agreement with the presence of esters as was suspected from the infrared, mass, and proton NMR spectra. The five acetyl methyl carbon atoms lead to signals in the very narrow range of $\delta = 20.3$ to $\delta = 20.8$. There are five more signals in the region $\delta = 61.4$ to $\delta = 89.0$. The DEPT spectra indicate one CH_2 and four CH groups and thus the presence of six protons bonded to these carbon atoms. In the proton NMR spectrum, however, besides the methyl groups seven protons were detected. Since there is no indication for an X–H group in the infrared spectrum, we must conclude that two methine carbon atoms have the same chemical shift. The high intensity of the signal at $\delta = 69.7$ is in perfect agreement with this interpretation. The chemical shift of $\delta = 89.0$ together with a proton chemical shift of $\delta = 6.33$ shows that one methine group must be attached to two oxygen atoms.

15.2 Structural Assembly

The following structural fragments have been found so far:

5 CH_3–COO–

1 CH_2–(O)–

4 CH–(O)–

1 (O)–CH–(O)

We need at least one ether oxygen atom to accommodate the above fragments. The molecular formula becomes thereby $C_{16}H_{22}O_{11}$ (mass 390). Since it indicates six double bond equivalents while only five carbonyl groups and no C=C double bonds are present, the molecule must contain one ring.

Further specifications can be made on the basis of the proton NMR spectrum. We realize that the high-field part of it (signals at 4.1 ppm) belong to a strongly coupled spin system, so that the following first order interpretation is only approximate and needs confirmation by a computer simulation using the parameters obtained by first order interpretation. The signal at 6.33 ppm is a doublet with a splitting of 3.7 Hz. The same coupling is only found in the signal group at 5.1-5.2 ppm. According to the integration two protons absorb in this region. A closer inspection reveals a doublet of doublets at the high-field side (coupling constants 3.7 and ca. 10 Hz) and three lines for the low-field part with a splitting of approximately 10 Hz. The two coupling constants must be slightly different because the center line is broadened and thus does not have an intensity twice as large as the outer lines. The two lines at extreme position of these multiplets overlap and produce the most intense line at 5.12 ppm. The second coupling partner of the proton at 5.10 ppm must be the one at δ = 5.47 as shown by the relative intensities of the lines. The intensity difference demands for a coupling partner at lower field. There are two possibilities, the proton at 5.14 or the one at 5.47 ppm. However, if it were the one at 5.14, the intensity differences would be much larger due to second order effects. Both protons at 5.47 and at 5.14 ppm exhibit two coupling partners, with coupling constants close to 10 Hz.

The pieces of information found so far can be summarized as follows:

CH – CH – CH – CH – (CH)
6.33 5.10 5.47 5.14 ppm

We have not yet assigned the signal of the last methine group (in parentheses) and that of the methylene group. The three protons absorb at 4.27 (1 H) and ca. 4.1 ppm (2 H) and cannot be fully understood on the basis of first order rules. However, it is obvious that one of the methylene protons has its chemical shift at 4.27 ppm. It has a large (geminal) coupling of (-)13 Hz and a vicinal one to the CH which seems to be at near 4.12 ppm whereas the second methylene proton has a chemical shift of δ = ca. 4.1.

This information is summarized in the following scheme:

ca 3.7 Hz ca. 10 Hz ca. 10 Hz ca. 10 Hz ca. 4 Hz and ? ca. 13 Hz

—O CH—CH—CH—CH—CH—CH_2
—O O O O O O

6.33 5.10 5.47 5.14 4.12 4.27/4.1 ppm

These coupling constants imply a six-membered ring in a chair conformation.

Check it in SpecTool:
Values of coupling constants can be found on the pages $\mathbf{J_{geminal}}$ and $\mathbf{J_{vicinal}}$ under **HMNR Data**, **Special**.

Vicinal coupling constants of ca. 10 Hz are only observed for diaxial position of the coupling partners. Also the chemical shifts fit to this interpretation. One methine proton is at 4.12 ppm, about 1 ppm at higher field than the others so that this CH group must be bonded to an ether oxygen instead of an ester.

Thus we have derived both the configuration and the conformation of the compound which is pentaacetyl α-glucose:

H OAc H O AcO AcO H H H OAc OAc

15.3 Comments

15.3.1 Mass Spectrum

The argument used in the discussion of the mass spectrum, implying that a mass difference of 16 mass units is unlikely from the molecular ion, needs some qualification. Such a difference is chemically *a priori* not unreasonable, because methane and oxygen are possible neutral leaving groups. However, the probability of loss of methane without accompanying loss of methyl (mass difference 15) is negligible, while loss of oxygen is restricted to a few types of functional groups and is then either accompanied by loss of

17 mass units (OH) like in some N-oxides, epoxides and sulfoxides or by loss of 30 mass units as in nitro compounds. Since none of these side reactions occur, the original argument is applicable.

15.3.2 Infrared Spectrum

This compound exhibits C–H stretching vibrations just above 3000 cm^{-1}, seemingly indicating hydrogen atoms bonded to sp^2 hybridized carbon atoms, or a three-membered ring. In this case, it is the solvent absorption that mimicks a high frequency. There are, however, other structural elements that may exhibit C–H stretching vibration frequencies just above 3000 cm^{-1}, e.g., hydrogen atoms bonded to carbon atoms substituted with halogen atoms.

15.3.3 Proton NMR Spectrum

Since parts of the spectrum exhibit strong couplings the interpretation must be confirmed by simulation of the spin system [1].

15.4 References

[1] U. Weber, H. Thiele, G. Hägele, NMR Spectroscopy: Modern Spectral Analysis, VCH, Weinheim, 1997.

16 SpecTool

16.1 What is SpecTool?

SpecTool is a Hypermedia application. It contains reference data, reference spectra, and computational tools needed for the interpretation of molecular spectra. The information units are interconnected by a network of links. SpecTool has been developed to support the interpretation of molecular spectra. It is not intended for novices in molecular spectroscopy, since basic knowledge in this field is a prerequisite when using it.

SpecTool is neither an expert system nor a database.

Expert systems attempt to map the present state of knowledge by collecting basic facts and, in particular, formulating all relations between these facts in a comprehensive set of rules. The inference machine inherent in every expert system then attempts to infer conclusions about the problem at hand by combining the rules and facts, and presents the user with a completed solution. SpecTool also contains a comprehensive collection of basic facts related to the covered methods of molecular spectroscopy. However, it limits itself to presenting the user with those rules relevant to the given context and lets him decide. SpecTool never takes any decisions on its own behind the back of the user.

Spectroscopic databases attempt to document the spectral data for as many compounds as possible. It is then up to the user to identify and retrieve the data records relevant to the problem at hand. SpecTool, however, does not aim at indiscriminate universality, it rather limits itself to reference data for carefully selected prototype molecules. These references are then presented to the user in such a way that generalizations for a the current context are easily possible.

16.2 What is Hypermedia

The classical medium for storing data is the book. Its information is physically organized in blocks, i.e., pages, each of them having, in a natural linear order, exactly one previous and one following page. The logical structure of the information compellingly follows the clear and simple organization along the given linear sequence. The index of a book offers another linear structure that provides access points to the main information stream. In case of encyclopedias, the linear physical organization does not correspond to the logical one. The reader may jump from item to item according to the cross references. The sequence of blocks is only relevant for supporting the search. Similarly, in a classical database the sequence of the information blocks becomes irrelevant and two succeeding blocks need not be logically related to each other. Thus, in contrast to books, browsing through a data bank is hardly ever feasible. On the other hand, the number and types of indexes is almost unlimited.

Hypermedia try to combine the strong points of both approaches. Their logical structure is the closest to that of encyclopedias. A database-like structure allows for quick and easy access to any information block via classical indexing functions. This is of advantage if the user knows exactly what he is looking for. On the other hand, each information block is linked to all other logically related blocks. This allows the user to browse through the data compilation the same as through a book. It depends on the context which other blocks are logically related to a given one. Thus, there are many different pathways leading through the data base, emphasizing different facets, views, and problems. A hypermedia network, therefore, is like a series of books presenting the same basic data along different lines of thoughts. Of course, information is not limited to text and data may include programs, graphics, animated graphics, as well as sounds.

A hypermedia network thus consists of facts, tools for manipulating them, and connections between logically related clusters of facts and tools. The clusters of facts and/or data are referred to as nodes and the logical connections between them are called links.

16.3 Contents of SpecTool

In SpecTool, the node contents cover the spectroscopic techniques most frequently used for structure elucidation of organic compounds, i.e., mass spectrometry, ^{1}H- and ^{13}C-NMR spectroscopy as well as IR and UV/vis spectroscopy. The links implemented in SpecTool map the thought patterns of a chemist interpreting the spectra. The physical data structure is selected so as to render this mapping logical and easy.

ToolBook has been used to realize the system on personal computers under Windows. ToolBook can handle textual and pictorial information in various ways. Its interpreter language is suitable for developing versatile navigation tools as well as consistent data representations. Furthermore, it is appropriate for performing simple calculations. For more complex calculations, custom routines can be implemented. Links to stand-alone compiled programs are also available, allowing extra modules to be offered.

16.4 Basic Organization Units

The organization metaphors of ToolBook are pages which are collected in books. A page may contain printed and/or graphic information and have one or more selected areas which, when clicked at with the mouse, change the information displayed on the page, jump to any other page in any book, or start an internal or external program. These features allow a versatile data presentation, navigation structures, and the performance of simple to complex calculations.

16.5 Links

Links are implemented between nodes which contain logically related data. For clarity, one set of links have been organized in hierarchical tree-like structures. Additional links may exist between nodes, possibly on different hierarchical levels. The logical base of the link tree is a hierarchical classification scheme for organic compounds, taking into account the peculiarities of the respective spectroscopic methods. Each piece of data can be classified according to three mutually orthogonal points of views. It refers to a specified compound class, to a given spectroscopic method, and to a given data type, namely numerical ones (i.e., interpreted data, possibly generalized), reference spectra, spectra of the problems discussed in this book, or computational tools. Thus, the knowledge pool spans a three-dimensional hyperspace, its three axes being the compound type, the spectroscopic method, and the type of information. Each unit box in this scheme may again contain a hierarchical tree connecting the individual compounds (or compound subclasses) comprehended in the respective node. Hence, the information space has more than three dimensions and is, therefore, referred to as hyperspace.

The first level of the compound classification scheme is the same for all methods implemented and is represented in matrix form on the SpecTool Top. The columns represent the spectroscopic methods, the rows the data types. The selection buttons in the matrix expand the respective node so as to allow selection on the next hierarchical level, which may be different for different data types and different spectroscopic methods. Selections on the topmost level can be made from buttons on the Top Page, via a menu, or via a palette. On lower levels, where a meaningful generalization over compound classes, data types, and spectroscopic methods is not generally possible, navigation is done predominantly via specialized buttons.

16.6 Limitations of the Current Implementation and Known Problems

Using a commercial shell to implement a hypermedia system has many obvious advantages. However, in addition to the limitations imposed by the computer's operating system, it adds some additional constraints to the capabilities and performance. For the present version of SpecTool, which runs under ToolBook 1.5 US optimized for Windows 3.1 there are the following important limitations.

ToolBook (and in many aspects Windows) does not take into account the widely different screen resolutions of different monitors and graphic cards. SpecTool uses bitmap graphics and is designed to present itself reasonably nice on simple 640×480 screens, and at the same time to be compatible with high resolution systems. It is optimized for a screen resolution of 800×600. However, on very high resolution systems it presents itself on the monitor in almost postage stamp size. A workaround for high resolution systems is to have batch file that changes the screen resolution before calling SpecTool. This is probably not worthwhile for 1024×768 systems, but should be done with resolutions of 1280×1024 and higher.

SpecTool requires some non-standard characters (e.g. the superscript dot and plus sign to mark a radical-cation in mass spectrometry). Thus it uses its own special font, in which otherwise unused characters are associated with the special symbols. If the SpecTool font is not available, the operating system will substitute a font similar in size and style. However, this substitute font will show the replaced characters instead of the SpecTool symbols. Thus, before SpecTool is used, its special font has to be installed, using the standard Windows procedure for font installation.

When exporting data from a SpecTool page to a printer a bitmap representation should be used. Otherwise, the operating system will transfer text by sending ASCII codes to the printer, which will then print out the standard characters rather than the special SpecTool symbols. No problems arise when screen shots are printed out.

SpecTool proper and the 3 palettes run as independent window applications communicating via DDE. Thus, when using a palette to access a new book, the palette remains the active application, and mouse clicks in the main window may be ignored. A non-highlighted title bar in the main window characterizes this state. Just single click into the title bar to make the main application active. If SpecTool should run together with other large applications, hide the palettes: This sets free a significant amount of memory.

If changing the spectroscopic method at a point far down on the hierarchic tree on a machine not generously equipped with memory, the system may report an insufficient memory error or complain about not having enough resources and/or graphic memory. In this case just click on the OK button and continue by finishing the intended jumps manually.

SpecTool uses ToolBook's mechanism to call external applications. External applications are called asynchronously, the called application as well as the calling instance of SpecTool running simultaneously. It is thus possible to make SpecTool active again without closing the external application. This does not pose fundamental problems, but the amount of free resources will soon drop to zero. Thus, always close external applications before going back to SpecTool.

SpecTool 2.1 (and ToolBook 1.5) have been tested on Windows 95 and on Windows NT. There seem to be no incompatibilities, but on Windows 95 the building up of the screen is sometimes too slow, so parts of the previous screen remain visible inside the SpecTool window. In this case force a repaint by minimizing and then restoring the window (moving the window on the screen does not trigger a full repaint).

In addition to these technical limitations the version of SpecTool distributed with this book limits the access in the data and spectra section to those nodes which are relevant in the context of the compounds dealt with in this book. Furthermore, the external NMR prediction tools are limited to structures having the same empirical formula as the book's example compounds. Finally, the SpecLib section, which in the full version contains several thousand reference spectra, here contains the spectra of the compounds treated in the book.

17 Mass Spectrometry: Additional Remarks

17.1 Presentation of Data

Mass spectra were recorded on a VG-Tribid or a Hewlett Packard 5970 mass spectrometer and are presented as linear line graphs with mass numbered intensity maxima, normalized to the strongest peak (base peak) of the spectrum. This is the most convenient summary of the data for a quick survey and makes characteristic features like intensity distribution, ion series or isotope pattern easily perceptible, thereby outweighing the disadvantages against tabulated spectra of less accurate intensity specification and loss of weak signals.

17.2 Degree of Unsaturation. Calculation of Number of Double Bond Equivalents

The elemental composition of a saturated hydrocarbon is C_nH_{2n+2}. Incorporation of k double bonds or rings is tantamount to removal of $2k$ hydrogen atoms from that formula. A triple bond is equivalent to two double bonds. Halogen atoms can replace hydrogen. Thus F, Cl, Br, and I are hydrogen equivalents, their number is added to the number of hydrogen atoms when calculating the degree of unsaturation. Si, Ge, Sn, Pb are in the same sense carbon equivalents. Their number is added to the number of carbon atoms. Insertion of O (or of the oxygen equivalents S, Se, Te in their divalent form) does not alter the carbon-hydrogen ratio in saturated systems. They can be neglected in these calculations. Insertion of m trivalent N atoms (or of the nitrogen equivalents P, As, Sb, Bi) requires addition of m hydrogen atoms to maintain saturation, such as to make the general formula for saturated systems $C_nN_mH_{2n+2+m}$, where N stands for any one of the nitrogen equivalents. All other atoms are preferably evaluated by application of the general formula (eq 17.2) on the next page.

The number of double bond equivalents (or degree of unsaturation) F corresponds to the difference between the number of required hydrogen atoms for complete saturation $(2n + 2 + m)$ and the actual number x (including hydrogen equivalents) divided by two:

$$F = \frac{(2n + 2 + m) - x}{2} \qquad (17.1)$$

where n represents the number of carbon atoms plus carbon equivalents m the number of nitrogen atoms plus nitrogen equivalents x the number of hydrogen atoms plus hydrogen equivalents.

In a simplified approach, which suffices for the most common cases, that is for compounds containing only C, H, O, trivalent N, divalent S and the halogens, one can reduce the elemental composition formula to a hydrocarbon equivalent and calculate the degree of unsaturation according to the following steps:

1. omit oxygen and sulfur,
2. replace all halogens by hydrogens,
3. replace all nitrogens by CH groups,
4. compare the resulting hydrocarbon composition C_nH_x with the composition of a saturated hydrocarbon C_nH_{2n+2}. The number of unsaturations is given by half the number of missing hydrogens.

A comprehensive approach of general applicability is the calculation of the number of double bond equivalents by the following equation:

$$F = \frac{2 + \sum_i n_i(v_i - 2)}{2} \tag{17.2}$$

where n_i is the number of individual elements and v_i their respective number of formal valencies.

Even electron fragment ions in the mass spectrum (odd mass ionic species if they contain no or an even number of nitrogen atoms) are species with one open valence. A saturated system of this kind contains, therefore, one hydrogen less than specified above. This should be taken into account when calculating the degree of unsaturation in fragments.

If the nominal mass of the molecular ion is even, the number of hydrogen atoms is even also, unless an uneven number of hydrogen equivalents is present and vice versa.

Check it in SpecTool

The **MOLFORM** program (an **MS Tool**), which generates all molecular formulas compatible with the mass range and other element constraints, automatically calculates the number of double bond equivalents. Click on a formula in the generated list to see its number of double bond equivalents.

17.3 General Information from Mass Spectra

The intensity distribution of the ion signals in mass spectra reflects to some extent stability features of the investigated structures. Concentration of the overall ion yield in the molecular ion region indicates compact molecular arrangement as in purely aromatic compounds, largely conjugated or unsaturated polycyclic systems and the like, while in saturated aliphatic compounds the low mass range carries most of the total ion yield. Stable entities substituted by a few easily removable residues produce a few significant intensity maxima as common sense would suggest.

Sequences of intensity maxima (ion series) in the lower mass range and their respective m/z values are indicators of structural type and degree of saturation, which constitute valuable initial information for interpretation. The series 15 + (14)*n* (m/z 29, 43, 57, 71... "alkyl series") of elemental composition C_nH_{2n+1} or $C_nH_{2n-1}O$ is typical of saturated aliphatic hydrocarbons or ketone and aldehyde compounds or residues, the series 13 + (14)*n* (m/z 27, 41, 55, 69... "alkenyl series") indicates one double bond equivalent as in alkenes, cycloalkanes and cycloalkanones or monofunctionalized compounds, which easily eliminate a neutral molecule (e.g., alcohols lose water). Aromatic hydrocarbon residues result by degradation in the highly unsaturated "aromatic series" m/z 39, 51 ± 1, 64 ± 1, 78 ± 1, 91.... Singly bonded oxygen in saturated systems gives rise to the "oxygen series" m/z 31, 45, 59, 73□, nitrogen in aliphatic saturated compounds to the "nitrogen series" m/z 30, 44, 58, 72..., sulfur in saturated residues to the "sulfur or 2 oxygen series" m/z 47, 61, 75, 89.... Polycyclic saturated systems cause a gradual switching of sequences of maxima into more and more unsaturated series along the way up the mass scale. In general, the intensity distribution within such series is steadily rising or falling. Striking intensity jumps of individual members within a series (positive or negative) are always structurally significant and should be interpreted. If isolated intensity maxima are observed in the lower mass range, it is usually rewarding to determine, which ion series they belong to and consider their possible elemental composition. Even mass maxima within uneven series, and *vice versa*, are also always diagnostically important features.

Check it in SpecTool
The MS Tool **Homologous mass series** gives easy access to various ion series data.

17.4 Evidence for Elemental Composition from Isotope Peak Intensities

Most of the elements constituting organic molecules occur in nature as mixtures of different isotopic species. Their natural relative abundance is not constant in a strict quantitative sense, but constant enough to be characteristic of the individual element within the limits of accuracy of intensity measurements in normal qualitative analysis by mass spectrometry (standard deviations significantly less then 1% from the mean).

Check it in SpecTool
The MS Tool **Isotope Table** gives the isotope distribution of all elements, accessible by name of the element, by mass, or by number of isotopes.

Since mass differences between particular isotopes are very close to full mass units or multiples thereof, ions of a given elemental composition will always yield more than one signal in the mass spectrum. Species containing heavy isotopes are separated in the analyzer and give rise to isotope peaks, the intensity of which is related to type and number of individual atoms in the molecular formula by their respective natural isotope abundances. Among the most common elements in organic chemistry, F, P, and I are monoisotopic, and the natural abundance of heavy isotopes in H, N and O is too low to produce isotope peaks of sufficient intensity. In contrast, C contains enough ^{13}C in natural abundance to be significant, each carbon atom contributing about 1.1% to the isotope peak intensity at the next integer mass value of singly charged ions. Consequently, the intensity of a signal in % of the signal intensity at the previous integer mass value divided by 1.1 indicates the upper limit for the number of carbon atoms which can be present in the respective ionic species. The intensity of the first isotope peak can be higher than required by the number of carbon atoms present, due to protonation, which is fairly common, or due to contributions from other elements with appreciable natural abundance of heavy isotopes, e.g. Si (^{29}Si of 5.1 % abundance relative to ^{28}Si), but it cannot be lower.

In the second isotope peak the contribution due to the common elements C, H, N, O (i.e., the probability of an ion containing two ^{13}C, ^{2}H, ^{15}N, ^{17}O, or one ^{18}O) can be assumed to be approximately 10% of the intensity of the first ^{13}C-isotope peak. This is a very rough approximation indeed, but sufficient for purposes of qualitative analysis (deviation less than 1 % relative to $M^{+\cdot}$). If the first isotope peak contains significant amounts of protonated species without ^{13}C, the corresponding ^{13}C-isotope intensity must be taken into account in the second isotope peak. Intensities of second isotope peaks in excess of these values indicate elements with higher natural abundances of heavy isotopes two mass units apart such as Si (^{30}Si of 3.9 % abundance relative to ^{28}Si), S (^{34}S of 4.4 % abundance relative to ^{32}S), Cl (^{37}Cl of 32.2 % abundance relative to ^{35}Cl), Br (^{81}Br of 98 % abundance relative to ^{79}Br) or elements occurring less frequently in organic compounds. The presence of several such elements gives rise to characteristic peak clusters, the intensity distribution of which can be calculated from natural abundances.

Check it in SpecTool

In practice, the most important clusters of this type are those due to combinations of Cl and Br which are displayed in the **MS Data** section, **Halogen compounds**.

Isotope clusters for arbitrary combination of elements can be calculated with the MS Tool **Isotope Pattern**.

17.5 Evidence for Elemental Composition in Low Resolution Mass Spectra

Structurally non-specific information concerning elemental composition or presence or absence of specific heteroatoms can be drawn not only from characteristic isotope peak intensities but also from characteristic mass values of fragment peaks or of mass differences between molecular ion and fragments.

Check it in SpecTool

Indicators for heteroatoms (ion series, key fragments and neutral losses) are given in the **MS Tools** section.

17.6 High Resolution Data

If elemental composition assignments are based on accurately measured mass values, two limiting conditions must always be kept in mind:

1. The correlation between accurate mass and elemental composition is significant only within the limits of resolving power achieved in the measurement and provided that the assumption of a uniform elemental composition of the measured ion beam is correct. In an unresolved multiplet, the measured mass value will represent the weighted average of all components in the respective peak and its correlation with one specific elemental composition is bound to fail, if the standard deviation of the of mass measurement is smaller than the mass difference between individual compound masses and weighted average.
2. The correlation between accurate mass value and elemental composition is naturally ambiguous at higher masses, if many hetero atoms must be allowed for, irrespective of the accuracy of measurement. There is always more than one possible combination of elements resulting in the same accurate mass, the number increasing with increasing number and type of atoms to be considered. An appropriate choice among these can only be made by controlling the respective isotope peak intensities and masses, by consistency checks over wide ranges of the spectrum and/or by including additional information (like other spectroscopic data or results of combustion analysis) as selective arguments.

Check it in SpecTool
The **MOLFORM** tool lists all elemental compositions compatible with a given mass, using a variety of user-specified constraints, including mass range, number and type of atoms, and weight percentages.

17.7 Impurities in Mass Spectra

No sample is ever absolutely pure. The most common impurities are traces of solvents, or phthalates from plasticizers in commercial polymers or from pump fluids, or traces of homologous compounds which are usually contained in reagents as well as natural products. Other impurities may, of course, be present in many cases as well and complicate the interpretation. Their spectra overlay those of the main compound and give rise to ambiguities unless they are identified as not belonging to the subject of analysis.

The nature of sample admission techniques and the high sensitivity inherent in mass spectrometry can lead to spectra which do not reflect the relative amount of impurity. If the sample is not completely evaporated and admitted from a reservoir, fractionation effects may occur, the extent of which depends on the difference in vapor pressure between compound and impurity at the given temperature. In individual scans enrichments by factors of thousands may result and preclude even a qualitative estimation of the amount of impurity, if differences in volatility are large. Consequently there is usually little point in trying to verify presumed impurities seen in mass spectra in other spectroscopic data, unless comparable volatilities (as in homologous compounds) can reasonably be assumed. Signals due to impurities in the molecular ion region can interfere with a correct determination of molecular mass, because they may mimic chemically unreasonable mass differences and thereby suggest higher values than are really involved. If ambiguity cannot be eliminated by considering mass differences and establishment of two independent degradation series, the spectrum needs to be rerun and fractionation verified by comparing spectra obtained at different temperatures.

Identification of the correct molecular mass or even the correct elemental composition is by the same token not a reliable indication for sample purity, because impurities may have been fractionated away before the specific spectrum was registered, or they may be of lower mass, or they may be insufficiently volatile under the given experimental conditions and thereby be impossible to detect, or they may simply be isomers.

18 Infrared Spectroscopy: Additional Remarks

18.1 Presentation of Data

The infrared spectra were recorded on a Perkin-Elmer 1600 series Fourier transform spectrophotometer and are presented as plots of transmittance versus frequency. The frequency scale is inverted, the frequencies increasing from right to left. Frequency values are given as wave numbers ν in cm^{-1}. There is a change of scale at $\nu = 2000$ cm^{-1}. The scale to the right of this value, corresponding to lower frequencies, is expanded by a factor of 2 relative to the scale to the left of the 2000 cm^{-1} point representing higher frequency values. The matrix used for recording the infrared spectra is given at the bottom.

18.2 Prediction of Infrared Stretching Frequencies

A rough prediction of IR stretching frequencies can be made using the harmonic oscillator model. The absorption frequency of the stretching mode of a two-atom oscillator depends primarily on the mass of the two atoms involved and on the force constant of the bond between them. It may be estimated according to the following equation:

$$\nu_{st}[\mathrm{cm}^{-1}] = 1303\sqrt{k\left(\frac{1}{m_1} + \frac{1}{m_2}\right)} \qquad (18.1)$$

where m_1 and m_2 are the relative atomic masses; k is a factor characterizing the type of bond that can assume values of 5, 10, and 15 for single, double, and triple bonds, respectively. Due to the many simplifying assumptions (e.g., harmonic oscillator that is fully independent from the remainder of the molecule), predicted values give the order of magnitude only. The equation is, however, useful for predicting isotope shifts, as exemplified in the following.

Predicted for C–H st:

$$\nu_{C-H} = 1303\sqrt{5\left(\frac{1}{12} + \frac{1}{1}\right)} = 3033\,\mathrm{cm}^{-1} \qquad (18.2)$$

Prediction of C–D st:

$$\nu_{C-D} = 1303\sqrt{5\left(\frac{1}{12}+\frac{1}{2}\right)} = 2225\,\mathrm{cm}^{-1} \qquad (18.3)$$

Thus, C–D stretching frequencies in deuterated hydrocarbons are expected around 2200 cm^{-1}.

If the stretching vibration frequency ν_o of the unlabelled compound is known, the respective frequency ν_L in the labelled compound can be predicted with considerable accuracy by using the following relation:

$$\nu_L = \nu_o\sqrt{\frac{\frac{1}{m_1}+\frac{1}{m'_2}}{\frac{1}{m_1}+\frac{1}{m_2}}} \qquad (18.4)$$

where m_1, m_2, and m'_2 are the relative atomic masses of the two atoms involved and of the isotope.

As an example the D–O stretching frequency in liquid D_2O is calculated from the H–O stretching frequency in H_2O (3490 cm^{-1}) as:

$$\nu_{D\pm O} = 3490\mathrm{cm}^{-1}\sqrt{\frac{\frac{1}{16}+\frac{1}{2}}{\frac{1}{16}+\frac{1}{1}}} = 2539\mathrm{cm}^{-1} \qquad (18.5)$$

which compares favorably with the observed value of 2540 cm^{-1}.

18.3 Overtones, Combination Bands, Fermi Resonance

In general, infrared absorption bands correspond to transitions from the ground state to the first vibrationally excited state. However, in some cases the vibration quantum number may change by 2. The corresponding absorption bands, the so-called overtones, are found at frequencies equal to approximately twice the frequency of the fundamental band. Furthermore, the absorption of infrared radiation may cause the simultaneous excitation of two or even three vibrational modes, giving rise to combination bands. These are found at frequencies corresponding to the sum (the molecule gains energy in both modes) or to the difference (the molecule gains energy in one mode, but loses a smaller amount in the other mode) of the fundamental frequencies involved.

Combination bands and overtones exhibit generally much lower absorption intensities than fundamental vibrations. Nevertheless, they can sometimes be of considerable diagnostic value. For instance the series of overtones and combination bands observed for benzene derivatives between 2000 cm^{-1} and 1600 cm^{-1} are useful for identification of the substitution pattern. Overtones of the out-of-plane deformation vibrations of the hydrogen atoms in terminal methylene groups, found near 1850 cm^{-1} to 1800 cm^{-1}, are helpful for the identification of this structural element.

It may happen that an overtone or combination band accidentally falls very close to a fundamental frequency. If both are of the same symmetry type, the two transitions will interact to give two new transitions, one of higher energy (higher frequency) and one of lower energy (lower frequency) than the original pair. In addition, the total intensity is redistributed between the two new transitions in such a way as to give two bands of similar absorption intensity. This type of interaction is termed Fermi resonance. Thus, a normally weak overtone or combination band may gain enough intensity to become an important absorption band. Some structural types show the phenomenon of Fermi resonance reliably in their spectra. An example are the chlorides of benzoic acids. Here, the carbonyl stretching frequency interacts with the overtone of a band near 875 cm^{-1}, giving rise to two bands in the carbonyl region near 1775 cm^{-1} and 1745 cm^{-1}. The two bands characteristic for aldehydes which are observed on the low frequency side of the C–H stretching region are most probably also due to Fermi resonance (interaction of the C–H stretching vibration with an overtone of the C–H deformation vibration).

18.4 Band Shapes and Intensities in Infrared Spectra

In contrast to spectroscopy in the ultraviolet and visible wave length region, it is not easily possible to give quantitative measures for absorption band intensities in infrared spectroscopy. The reason for this lies primarily in the fact that monochromacity and wavelength resolution of today's instruments is of the same order of magnitude as the bandwidth of infrared absorption bands, which strongly affects the band shape. Thus, in qualitative organic analytical chemistry the absorption intensity is generally expressed by subjective classifications, e.g., weak, medium, strong and very strong. In the regions where the transmission of the solvent is less than a few % the light energy reaching the detection system is insufficient for reliable operation. Here, the recording system behaves erratically, no interpretation of the spectra is possible, and the output is generally replaced by a horizontal line.

Not all vibrational modes give rise to an infrared absorption band. A prerequisite for a vibration to be infrared active is that the vibration must cause a change in the dipole moment. For example, the bending and antisymmetric stretching modes of CO_2 (which, of course, is a linear molecule) cause the molecule to have a dipole moment except in its equilibrium structure. Thus, both vibrations lead to an infrared absorption. In the symmetric stretching vibration the dipole moment does not change. Thus, there is no infrared absorption corresponding to this vibration, the symmetric stretching mode is not infrared active. Raman spectroscopy has different selection rules. For a vibration to be Raman active, a change in polarizability of the molecule during the vibration is required. For example, in the symmetric stretching vibration of CO_2, the electrons are further away from the nuclei as the bond stretches. Thus, they are less strongly attracted to the nuclei and are more polarizable. The inverse holds when the bond shortens.

Consequently, the symmetric stretching vibration of CO_2 is Raman active. In the bending vibration, bond length does not change, and in the antisymmetric stretching vibration the effects of both C=O bonds cancel each other. Thus, one would expect for CO_2 a Raman spectrum consisting of just one line. Accidentally, the frequency of this Raman band falls close to twice the frequency of the bending vibration (667 cm^{-1}). Thus Fermi resonance occurs, leading to two Raman peaks of similar intensity at 1388 and 1285 cm^{-1}.

18.5 Spurious Bands in Infrared Spectra

Traces of water in carbon tetrachloride and chloroform may give rise to bands near 3700 and 3600 cm^{-1} In addition, a weak broad band is observed around 1650 cm^{-1}. Water vapour exhibits many sharp bands between 2000 cm^{-1} and 1280 cm^{-1}. If present in relatively high concentrations, these bands may be the source of artifacts which are particularly annoying when scanning through a steep flank of a strong peak. Sometimes puzzling shoulders on carbonyl absorption bands can be explained by this effect.

Dissolved carbon dioxide exhibits an absorption band at 2325 cm^{-1}. In solutions containing amines, the dissolved carbon dioxide, together with the ubiquitous water, may form carbonates. Then the spectrum exhibits unexpected bands due to the protonated amine function. In the vapour phase, the absorption of carbon dioxide may be the cause of increased base line noise at 2360 and 2335 cm^{-1}, which is accompanied by another band at 667 cm^{-1}.

Commercial polymeric materials often contain phthalates as plasticisers, which sometimes find their way into "pure" analytical samples. They give rise to a band at 1725 cm^{-1}. Various chemical operations on the sample may transform the phthalates into phthalic anhydride which gives an absorption at 1755 cm^{-1}.

Another rather common impurity are the various silicones. These result usually in a band at 1265 cm^{-1}, together with a broad band in the 1100 cm^{-1} to 1000 cm^{-1} region. If carbon tetrachloride evaporates from a leaky cell, a band at 793 cm^{-1} is observed. However, for liquid carbon tetrachloride the respective band appears at 788 cm^{-1}. This band is observed if the outer wall of the cell is contaminated with carbon tetrachloride.

19 NMR Spectroscopy: Additional Remarks

19.1 Presentation of Spectra

19.1.1 Proton NMR Spectra

The original proton NMR spectra are directly reproduced throughout. They were generally recorded at 200, 400 or 500 MHz on a Bruker-Spectrospin DPX-200, AMX-400 or AMX2-500 spectrometer at room temperature using tetramethylsilane (TMS) as internal reference. In some cases, e.g., for spectra recorded in D_2O, an approximate referencing method was applied by using another sample containing the solvent and the reference compound (a water soluble derivative of TMS).

19.1.2 Carbon-13 NMR Spectra

The depicted proton wide-band decoupled spectra were recorded at 50, 100, or 125 MHz on a Bruker-Spectrospin DPX-200, AMX-400 or AMX2-500 spectrometer at room temperature using tetramethylsilane as internal reference. In some cases, the line of the solvent was used as an approximate reference. DEPT135 and DEPT90 spectra are given throughout for the determination of the number of attached protons to the individual carbon atoms. In ideal cases, i.e., perfect pulse widths and coupling constants close to $1/2\tau$ in DEPT135 spectra CH_3 and CH lead to positive signals and CH_2 to negative ones. Ideally, only CH groups are observed in DEPT90 spectra. In reality, the ideal conditions are often not fulfilled and additional small signals may be observed.

19.2 Rules for the Interpretation of Coupling Patterns

19.2.1 Definitions

It is useful to define isochronicity (chemical shift equivalence) and magnetic equivalence (chemical shift and coupling equivalence) as given below. The term "chemical equivalence" which is often used erroneously in this context should be avoided, because it is not precisely defined, and in several respects not directly relevant to NMR spectroscopy.

Isochronous nuclei

Nuclei are isochronous, if they do not have any measurable difference in their chemical shifts under the given experimental conditions. Isochronicity may be a consequence of molecular symmetry, fast intra- or intermolecular exchange or it can be purely accidental. Except for homotopic (see below) nuclei, isochronous nuclei may become non-isochronous under different experimental conditions.

Magnetic equivalence

Nuclei are magnetically equivalent if they are isochronous and if they exhibit identical coupling constants with every individual member of the set of all other nuclei. The members of an isolated set of isochronous nuclei which are not coupled to any other nuclei are also magnetically equivalent. When chemical shift and coupling equivalence is not a consequence of rotational symmetry, magnetically equivalent nuclei may become nonequivalent if the experimental conditions are changed.

Nomenclature of spin systems

A spin system extends over a set of nuclei which are connected by an unbroken coupling path. If there are two sets of nuclei in a molecule where no member of one group couples with any member of the other, the two sets can be treated as independent spin systems. Two different nomenclature systems are widely applied (cf. [1]), but only the one used in this volume is presented here.

A set of magnetically equivalent nuclei is denoted by a capital letter with an index giving the number of magnetically equivalent nuclei within the group. Isochronous nuclei which are not magnetically equivalent are denoted by the same capital letter, but are distinguished by primes. Nonisochronous nuclei are denoted by letters adjoining in the alphabet if they are strongly coupled so that higher order spectra result (chemical shift difference small relative to coupling constant, i.e. $\Delta\nu$ [Hz] < ca 10 J [Hz]). If the spins are weakly coupled (the chemical shift differences are large, i.e. $\Delta\nu$ [Hz] > ca. 10 J [Hz]), the nuclei are symbolized with letters far apart in the alphabet. Note that higher order spectra can be often simplified by suitably changing the experimental conditions, in particular by recording the spectrum at higher magnetic field strength. However, the presence of isochronous but magnetically nonequivalent nuclei with non zero coupling will always lead to higher order spectra (their chemical shift difference is zero by definition).

19.2.2 General Rules

The following rules apply both to first order and higher order spectra.

1. Coupling between magnetically equivalent nuclei does not affect the spectrum and is, therefore, not defined by the spectrum.
2. The coupling constants do not depend on the strength of the applied magnetic field. Since on the other hand, the chemical shift differences measured in Hertz depend on it, spin systems leading to higher order spectra at a given magnetic field may change to first order spectra at a higher magnetic field strength.

3. Each spin-spin interaction is mutual: if a nucleus *A* is coupled with another nucleus *B* (or *X*) than *B* (or *X*) is coupled with *A* to exactly the same extent: $J_{AB} = J_{BA}$.

19.2.3 First Order Spectra

First order spectra are observed, if the chemical shift differences between all magnetically nonequivalent nuclei within a spin system are large relative to the corresponding coupling constants. If there is no coupling between two sets of nuclei they constitute two independent spin systems. In practice, as a rule of thumb first order spectra may be expected if the following condition applies:

$$\Delta\nu_{ij} > \text{ca. } 10\ J_{ij} \qquad (19.1)$$

for all pairs of magnetically nonequivalent nuclei in a spin system. If a spin system exhibits a higher order part, the higher order effects will in general complicate the signals of the remaining nuclei as well.

For isochronous ($\Delta\nu_{ij} = 0$) but magnetically nonequivalent nuclei, higher order spectra are to be expected regardless of the value of the coupling constants between the isochronous nuclei (for example *AA'XX'* spin systems consist generally of 20 lines whereas first order rules would predict only 8 lines). Such a spin system can never be simplified by applying higher magnetic field strengths.

1. The multiplicity of the signal of *A* in an A_mX_n spin system is determined by the number *n* and spin quantum number *I* of the nuclei *X* by the relation $(2nI + 1)$. For protons, carbon-13 and other spin 1/2 nuclei this amounts to $(n + 1)$. If there are more than two interacting groups such as in an $A_nM_mX_p$ system of spin 1/2 nuclei, the multiplicity of the signal *A* will be given in general by $(m + 1)\ (p + 1)$. For special ratios of the coupling constants some lines may coincide, so that the number of observed lines may be reduced. If for example $J_{AM} = J_{AX}$ in the above case, $(m + p + 1)$ lines will be observed.
2. The intensities and positions of the lines are symmetric about the centre of the multiplet, which corresponds to the chemical shift.
3. If a multiplet is produced by coupling with a single group of equivalent nuclei, a multiplet of equidistant lines will be produced, the spacing of the lines is equal to the coupling constant J. If this group consists of *n* equivalent nuclei with $I = 1/2$, the relative intensities of the lines within the multiplet are given by the coefficients in the binomial expansion: 1 : 1 ($n = 1$); 1 : 2 : 1 ($n = 2$); 1 : 3 : 3 :1 ($n = 3$); 1 : 4 : 6 : 4 : 1 ($n = 4$); 1 : 5 : 10 : 10 : 5 : 1 ($n = 5$); 1 : 6 : 15 : 20 : 15 : 6 : 1 ($n = 6$); etc. The corresponding relative intensities obtained by coupling with *n* equivalent $I = 1$ nuclei are 1 : 1 : 1 ($n = 1$); 1 : 2 : 3 : 2 : 1 ($n = 2$); 1 : 3 : 6 : 7 : 6 : 3 : 1 ($n = 3$); etc.
4. It is often useful to rationalize the splitting and intensities by drawing the splitting diagrams. It is of no importance, in which order the individual couplings are considered. Thus the two representations shown in Fig. 19.1 for the *X*-part of a hypothetical A_2MX system are equivalent.

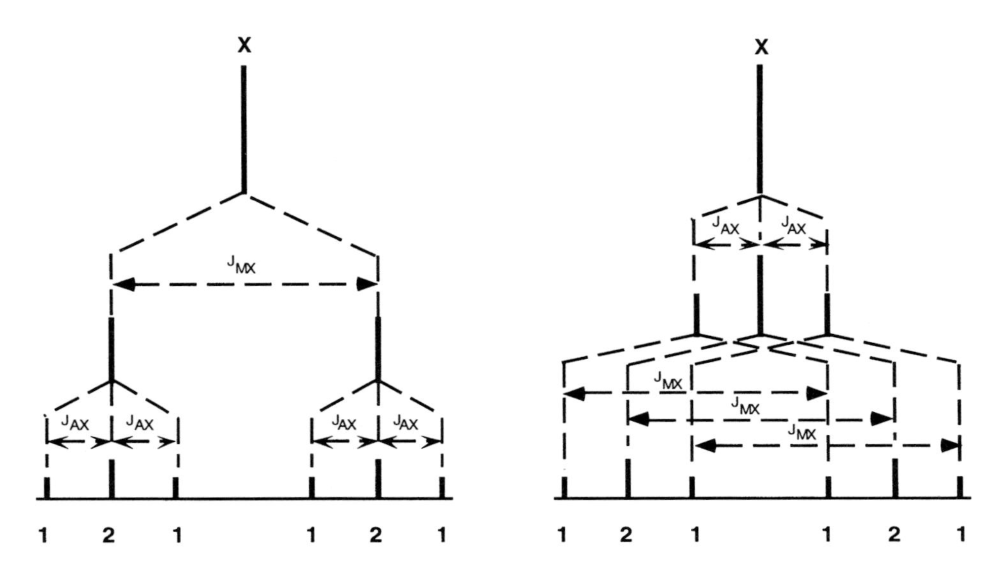

Fig. 19.1: Two equivalent ways of rationalization of the splitting for the *X*-part of a hypothetical A_2MX system

5. Since pure first order spectra only would exist if the ratio Δν/J were infinitely large, for homonuclear spin systems the first order analysis leads to minor errors. The splitting and positions of the lines are generally not affected if Δν/J > ca. 10 but their intensities are different from those predicted by first order rules. If for two lines the same intensities are predicted by first order rules, the one being closer to the corresponding coupling partner becomes more intense, the other one less intense than predicted. This effect is very useful in practice, because the connecting lines between peak tops are pointing in direction of the coupling partners (see Fig. 19.2).

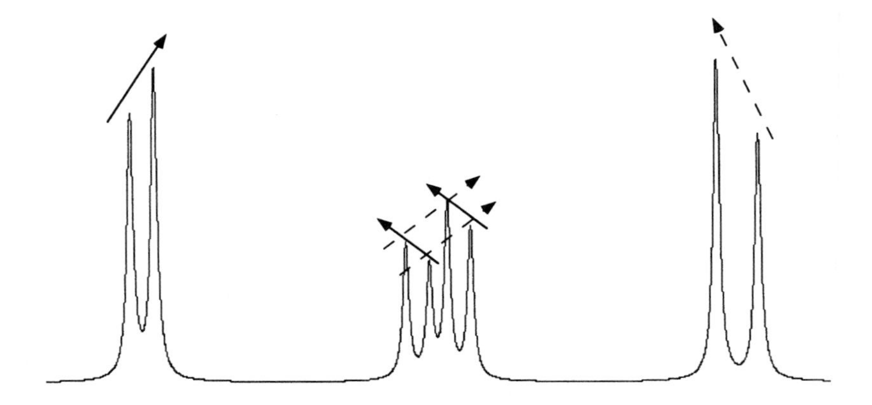

Fig. 19.2: Intensity differences between signals of a multiplet indicate the relative chemical shift of the corresponding coupling partner

19.2.4 Higher Order Spectra

If nuclei in a spin system are strongly coupled, i.e., if $\Delta\nu_{ij} < 10\ J_{ij}$, higher order spectra will be observed, which generally consist of more lines than predicted by rules for first order analysis. The set of parameters (chemical shifts and coupling constants) are, in general, not easily extracted from such spectra. If such higher order spectra are encountered, the chemist has the following possibilities for interpreting of the data:

1. Full analysis, i.e., determination of all available parameters by computer simulation of the spectra (see [1]).
2. Recording of the spectrum under changed experimental conditions such as spin decoupling, different solvents, higher magnetic field strength, isotopic substitution or application of shift reagents in order to obtain first order spectra.

In many cases qualitative features, i.e., information about the number of protons and the approximate chemical shift values, are sufficient for solving the problems, and no further analysis is necessary. Sometimes, misleadingly simple-looking spectra are observed ("deceptive simplicity") and interpretation according to first order rules leads to erroneous results. For example, *ABX* spin systems often lead to such errors. The *X*-part consists of six lines in the general case, but often only four lines are discernible. One may easily be tempted to interpret this part as a first order spectrum (*X* part of an *AMX* system) and assign the *AX* and *BX* coupling constants to the splittings. This leads however, in general, to false results, since the line splittings are not always equal to the coupling constants in an *ABX* spectrum. By the same token, even if only a triplet is observed for the *X* part, the coupling constants J_{AX} and J_{BX} may be different. However, if the *X* part consists of four lines, it positively indicates that $J_{AX} \neq J_{BX}$. In general δ_X and $|J_{AX}+J_{BX}|$ (distance between the outer lines in the four-line pattern) are directly accessible from the *X* part of the spectrum, whereas J_{AB} (the spacing found four times in the *AB* portion) is the only parameter directly obtainable from the *AB* part of *ABX* spectra.

Very useful qualitative information may be obtained from the presence or absence of symmetry in the spectra of four-spin systems. *AA'BB'*-, *AA'XX'*- and A_2B_2 (in practice rarely occurring) systems always give rise to a symmetrical pattern. The former two systems occur in benzene if it is *para* disubstituted with different substituents or *ortho* disubstituted with identical substituents. Although both types give rise to symmetrical patterns, the general shape of the spectra is not the same, so that a differentiation is generally possible. The spectra of *para* disubstituted benzenes have some resemblance of *AB* type spectra, while this is not true for *AA'BB'* systems obtained from *ortho* disubstituted benzenes with two identical substituents.

19.2.5 Computer Simulation

A set of *n* chemical shifts and *n*(*n*-1)/2 coupling constants unambiguously defines a spectrum. Several computer programs are available for calculation of spectra on the basis of exact solutions of quantum mechanical equations. Quite often approximate interpretation according to first order rules is possible but such an interpretation should be verified by computer simulation. Another volume of this series gives an in-depth presentation of the topic and uses a series of spectra from this volume for illustration [1].

19.3 Signal Intensities in Carbon-13 NMR Spectra

If a system is in thermal equilibrium, the signal intensities in NMR spectra are proportional to the number of absorbing nuclei. Carbon-13 NMR spectra are, in general, recorded under experimental conditions which cause a non-equilibrium population of spin states. The most important factors influencing the line intensities in routine carbon-13 NMR spectra are discussed in this section.

19.3.1 Saturation

Carbon-13 NMR spectra are today exclusively recorded using the pulse Fourier-transform technique where a short and intense pulse is applied to excite all carbon-13 nuclei present. The response of the system, the interferogram, containing all frequencies present is then recorded. If a 90° pulse was applied, which leads to the maximum intensity following a single pulse, a nucleus reaches 99.98% of the equilibrium magnetization after a time of 5 T_1 s, where T_1 is the spin-lattice relaxation time of the nucleus. In practice, a great number of interferograms are recorded whereby the signal to noise ratio is improved by a factor of $\sqrt{n}$ if n interferograms are accumulated. The pulse rate is generally kept higher than necessary for complete relaxation of all nuclei present. The interferograms are thus recorded under conditions of partial saturation. Nuclei with longer relaxation times are more extensively saturated (their spin states are further away from the equilibrium) and give thus rise to reduced signal intensities while nuclei with the same relaxation time yield always the same signal intensity regardless of the extent of saturation.

With the exception of small molecules and of some quaternary carbon atoms in organic compounds, practically all carbon-13 nuclei relax by the way of dipole-dipole interactions with protons. The most important factors influencing the dipole-dipole relaxation time are the number of protons attached to a carbon atom and the rotational mobility of the C–H vectors. In most cases, the following sequence of relaxation times is observed:

$$CH_2 < CH < CH_3 << C \qquad (19.2)$$

The relaxation time of methylene carbon atoms is shorter than the one of methine carbon atoms because of the higher number of directly attached protons. Methyl carbon atoms exhibit, in general, longer relaxation times than methylene and methine carbon atoms as a consequence of their higher rotational mobility. Since the magnetic dipole-dipole interaction is dependent on the sixth power of the internuclear distance, quaternary carbon atoms exhibit much longer relaxation times than protonated carbon atoms. The sequence of expected signal intensities caused by partial saturation is thus:

$$CH_2 \geq CH \geq CH_3 \geq C \qquad (19.3)$$

Equal intensities are only expected if no saturation occurs and no other factors influence the line intensities. The relaxation times can be reduced by the application of paramagnetic relaxation reagents such as chromium acetylacetonate, ($Cr(acac)_3$).

19.3.2 Nuclear Overhauser Effect

The equilibrium population and, therefore, the line intensity of a nucleus can be influenced by irradiation at the Larmor frequency of another nucleus. Such effects, which occur in double resonance experiments, if the nuclei are coupled by dipolar interactions are termed nuclear Overhauser effect (NOE).

Since carbon-13 nuclei relax in most cases through dipolar interactions with protons, proton decoupling influences the carbon-13 line intensities. The intensities I increase thereby by a factor η which depends on the relaxation mechanism of the respective carbon-13 nucleus:

$$I = 1 + \eta = 1 + 2\frac{T_1}{T_{1\mathrm{CH}}} \qquad (19.4)$$

where $T_{1\mathrm{CH}}$ is the ^{13}C-^{1}H dipolar relaxation time and T_1 the overall spin-lattice relaxation time:

$$\frac{1}{T_1} = \frac{1}{\sum_{\mathrm{i}} T_{1\mathrm{i}}} \qquad (19.5)$$

where $T_{1\mathrm{i}}$ is the relaxation time due to the i-th mechanism.

If the proton-carbon-13 dipolar interaction is the only effective relaxation mechanism, T_1 becomes equal to $T_{1\mathrm{CH}}$ and the line intensity is three times higher than it would be without NOE. If on the other hand other mechanisms dominate, T_1 will be much smaller than $T_{1\mathrm{CH}}$ and no nuclear Overhauser enhancement occurs. Relaxation reagents, therefore, reduce the NOE.

The shorter the ^{13}C-^{1}H dipolar relaxation time the smaller is the probability of other mechanisms being significant. The order of intensity enhancements by NOE corresponds thus exactly to the sequence (19.3) given above for the expected line intensities due to partial saturation.

If full NOE occurs for all nuclei, the relative line intensities are not influenced. It is to be noted, that equation (19.3) only applies if the molecular rotational mobilities are high relative to the reciprocal of the resonance frequencies (extreme narrowing condition). This is practically always the case for small or medium sized organic molecules in common solvents at room temperature, if low or medium magnetic field strength is applied. At high magnetic field strength of 10 tesla or more (i.e., > ca. 100 MHz for ^{13}C) the extreme narrowing condition may not be met even for medium sized organic molecules. This leads to a reduced NOE even where the proton carbon-13 dipolar interaction is the predominant relaxation mechanism. Particular care must, therefore, be taken for the interpretation of line intensities in such cases.

Elimination of NOE can be easily achieved by applying the decoupling field in a pulsed mode. Since the multiplets collapse almost instantaneously when the decoupling field is turned on, whereas the NOE builds up with the time constant in the order of T_1, the NOE may be suppressed by gating the decoupler so that it is on only during each data acquisition period but off during each pulse delay.

19.3.3 Intensities of Solvent Signals

Generally, deuterated solvents are applied in carbon-13 NMR spectroscopy, since a deuterium reference signal is used for the stabilization (lock) of the magnetic field. The relaxation times of the carbon-13 nuclei in deuterated solvents are very long for several different reasons. First of all, the usual solvents are small and highly mobile molecules, leading to less effective dipolar interaction. Secondly, since dipolar relaxation with deuterium is much less effective than with proton, a further reduction of the contribution of this mechanism follows. The pulse rates usually applied thus cause extensive saturation of the solvent signals. Furthermore in nonprotonated molecules the NOE enhancement is absent. Finally, coupling with deuterium leads to line splittings and thereby to further reduction of the intensity of individual lines.

19.4 Influence of Molecular Symmetry and Conformational Equilibria on NMR Spectra

19.4.1 Introduction

The influence of symmetry and fast conformational equilibria on the NMR spectra very often causes difficulties in practical analysis which have led to numerous erroneous statements in the literature and even in textbooks. In this section we first present a consistent set of useful definitions and then give rules to predict the type of the expected spin system on the basis of symmetry considerations. Specific examples are then given to amplify the use of these rules. The prediction of the type of the expected spin system is often sufficient for spectra interpretation and it is also the first step if computer simulations are performed to evaluate the exact parameters [1].

19.4.2 Definitions

Relationships between pairs of nuclei of the same kind in a given structure

To predict isochronicity, the symmetry relation between two nuclei of the same kind is of relevance. Symmetry relations always refer to one specified structure, i.e., to fixed positions of all nuclei. Different conformations of a molecule define different structures.

Nuclei are *constitutionally equivalent* if they have the same connectivity (bondedness).

Nuclei are *diastereotopic* if they are constitutionally equivalent but not equivalent by symmetry.

Nuclei are *enantiotopic* or equivalent by reflection if the structure has an improper axis which interchanges the nuclei but no corresponding proper axis. Note that a symmetry plane and an inversion center are special cases of improper axes.

Nuclei are *homotopic* or equivalent by rotation if the structure exhibits a proper axis which interchanges the nuclei.

Relationships of one nucleus with a pair of nuclei of the same kind in a given structure

To predict the equivalence of two coupling constants, the symmetry of the coupling paths is of relevance. Although the coupling is mainly a through-bond interaction, for the present purpose the coupling path can be considered as a connecting line (edge) between the respective nuclei. To consider the relation of two nuclei to a selected coupling partner, we consider the relation of the corresponding connecting lines. Again the symmetry relation can be non-symmetric, symmetric by reflection, or symmetric by rotation. In the latter two cases the coupling partner must be located on the symmetry element (on the plane for improper symmetry operations or on the proper axis for proper rotations).

19.4.3 Influence of Symmetry Properties on the NMR Spectra

Homotopic nuclei are always isochronous and homotopic coupling paths always lead to equal coupling constants. In a non-chiral environment (including solvent, complexing agent, shift reagents, etc.), enantiotopic relationships induce equivalence of the chemical shifts or coupling constants. In a chiral environment, enantiotopic relationship does not imply magnetic equivalence. A chiral environment has, however, only a small effect on coupling constants. Thus loss of magnetic equivalence due to enantiotopic coupling paths in a chiral environment has not yet been observed experimentally. Of course, equivalence may occur by chance in any other case. For example, diastereotopic nuclei are often isochronous if the chiral part of the molecule is far away, e.g., if a chirality center is at a distance of at least 3 to 4 bonds.

19.4.4 Influence of Fast Equilibria

If chemical equilibria are fast, i.e. the mean life time of the species is small as compared to the relevant reciprocal chemical shift differences (measured in [Hz]) a single line with an average chemical shift is observed:

$$\delta = \sum_i \frac{c_i \delta_i}{c_{tot}} \tag{19.6}$$

where c_{tot} is the total concentration summed over all environments, c_i the relative concentration in the environment i, and δ_i is the respective chemical shift. The c_i are the concentrations of the nuclei, not of the species. Thus, in a 1 : 1 molar mixture of water and acetic acid, c_{water}: c_{COOH} is 2 : 1. An analogous relationship holds for the coupling constants in the case of a fast intramolecular exchange (but no couplings between the respective nuclei are observed if the fast exchange is intermolecular). By applying this equation, the parameters can be calculated, provided their values in the various environments as well as the concentrations are known. The application of this procedure for predicting the chemical shifts of a CH_3 and a CH_2 group in the hypothetical chiral compounds CH_3CXYZ (Fig. 19.3) and RCH_2CXYZ (Fig. 19.4; R, X, Y, Z are different substituents) is shown in the following paragraphs.

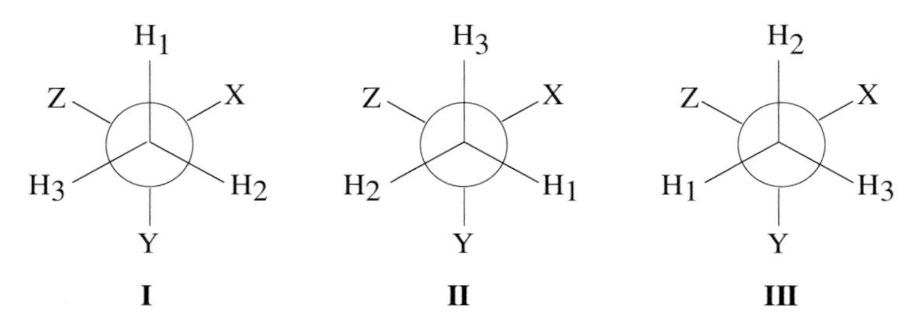

Fig. 19.3: Conformations considered for predicting the chemical shifts of the methyl protons in a hypothetical compound CH_3CXYZ

The chemical shifts of the three protons for fast exchange according to eq 19.5 are given in eqs 19.7 - 19.9 (c_i are the individual concentrations, c_{tot} ist their sum):

$$\delta_{H_1} = \frac{c_I}{c_{tot}} \delta_{H_1(I)} + \frac{c_{II}}{c_{tot}} \delta_{H_1(II)} + \frac{c_{III}}{c_{tot}} \delta_{H_1(III)} \tag{19.7}$$

$$\delta_{H_2} = \frac{c_I}{c_{tot}} \delta_{H_2(I)} + \frac{c_{II}}{c_{tot}} \delta_{H_2(II)} + \frac{c_{III}}{c_{tot}} \delta_{H_2(III)} \tag{19.8}$$

$$\delta_{H_3} = \frac{c_I}{c_{tot}} \delta_{H_3(I)} + \frac{c_{II}}{c_{tot}} \delta_{H_3(II)} + \frac{c_{III}}{c_{tot}} \delta_{H_3(III)} \tag{19.9}$$

The conformers **I** - **III** are in reality indistinguishable because they can be derived from each other by rotation of the methyl group around a bond, which corresponds to a threefold symmetry axis for it. As a consequence, eqs 19.10 - 19.13 are valid:

$$c_I = c_{II} = c_{III} = \frac{1}{3} c_{tot} \tag{19.10}$$

$$\delta_{H_1(I)} = \delta_{H_3(II)} = \delta_{H_2(III)} \tag{19.11}$$

$$\delta_{H_2(I)} = \delta_{H_1(II)} = \delta_{H_3(III)} \tag{19.12}$$

$$\delta_{H_3(I)} = \delta_{H_2(II)} = \delta_{H_1(III)} \tag{19.13}$$

By combining eqs 19.10 - 19.13 with eqs 19.7 - 19.9 it becomes evident that:

$$\delta_{H_1} = \delta_{H_2} = \delta_{H_3} \tag{19.14}$$

The situation is different for methylene protons:

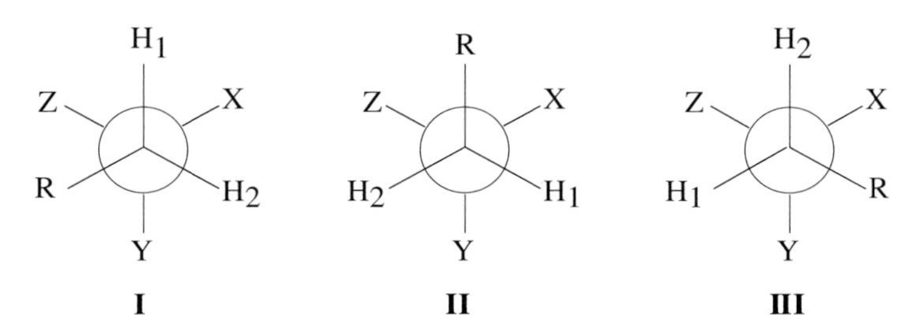

Fig. 19.4: Conformations considered for predicting the chemical shifts of the methylene protons in a hypothetical compound RCH_2CXYZ

The chemical shifts of the two protons for fast exchange according to eq 19.5 are given in eqs 19.15 - 19.16 (c_i are the individual concentrations, c_{tot} is their sum):

$$\delta_{H_1} = \frac{c_I}{c_{tot}} \delta_{H_1(I)} + \frac{c_{II}}{c_{tot}} \delta_{H_1(II)} + \frac{c_{III}}{c_{tot}} \delta_{H_1(III)} \quad (19.15)$$

$$\delta_{H_2} = \frac{c_I}{c_{tot}} \delta_{H_2(I)} + \frac{c_{II}}{c_{tot}} \delta_{H_2(II)} + \frac{c_{III}}{c_{tot}} \delta_{H_2(III)} \quad (19.16)$$

In the general case, the concentration of the individual conformers is not equal because the axis of rotation is not a local symmetry axis for the CH_2R group. Thus, the substituent R has different *gauche* substituents in each case:

$$c_I \neq c_{II} \neq c_{III} \quad (19.17)$$

Furthermore, there is no reason for an equivalence of chemical shifts of protons in virtually the same position (e.g., H_1 in **I** and H_2 in **III**). This can be easily seen, if one assumes for example a bulky substituent Y. The idealized conformations as shown in Fig 19.3 are not exactly valid: To reduce vicinal steric interactions the CH_2R group will be pushed clockwise in **I** and counterclockwise in **III** by the interaction of the bulky group with R so that the exact positions of H_1 in **I** and H_2 in **III** are different:

$$\delta_{H_1(I)} \neq \delta_{H_2(III)} \quad (19.18)$$

$$\delta_{H_2(I)} \neq \delta_{H_1(II)} \quad (19.19)$$

$$\delta_{H_1(III)} \neq \delta_{H_2(II)} \quad (19.20)$$

Thus, it is obvious that the chemical shifts of diastereotopic methylene protons cannot become equal due to fast rotation.

A treatment as shown above is, however, only rarely needed in practice. For fast conformational equilibria, the type of spectra can be predicted by applying the rules given below. They are, however, not valid for other kinds of fast exchange reactions.

For predicting the type of spectra in molecules with fast conformational equilibria, first atom groups are identified which rotate rapidly relative to the rest of a molecule around a bond, which is also symmetry axis for this group. All nuclei permutated by this rotation become isochronous and, with respect of all couplings to nuclei in the rest of the molecule, equivalent. Thus, the protons of a methyl group are always magnetically equivalent. The deeper reason for the validity of this rule is, as shown above, that due to the local symmetry both the population of the various conformers and the chemical shift of the different nuclei in the same position will be the same.

For the rest of the molecule with a fast conformational equilibrium, the rules described above for a rigid system may be applied to an arbitrary chosen conformation of the highest possible symmetry, regardless of whether this conformation is energetically reasonable. For example, the planar conformation of cyclohexane (!) is appropriate. The reason for this rule is that any other conformation will have a symmetrically equivalent counterpart, which will have the same population and therefore, the corresponding parameters will average out as if the molecule were in the single symmetrical conformation.

19.4.5 Examples

Molecule	Type of spectrum in non-chiral environment	Type of spectrum in chiral environment	Remarks
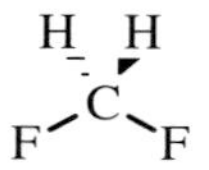 H H, C, F F	A_2X_2	$AA'XX'$	The ^1H and ^{19}F nuclei are pairwise homotopic, but the relevant coupling paths are only enantiotopic.
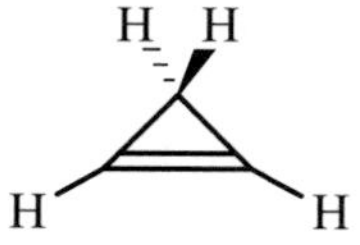 H H, H H (cyclopropene)	A_2X_2	$AA'XX'$	Topologically the same as the previous example.
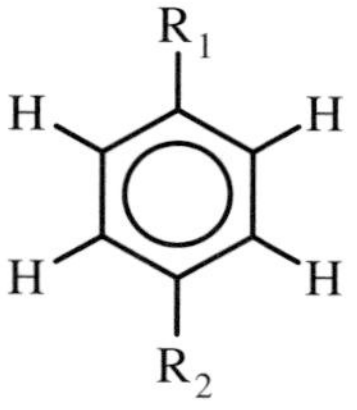 R1, H H, H H, R2 (benzene)	$AA'BB'$ or $AA'XX'$	$AA'BB'$ or $AA'XX'$	The protons are pairwise homotopic but the relevant coupling paths (i.e., one *ortho* and one *para* coupling) are not related by symmetry.
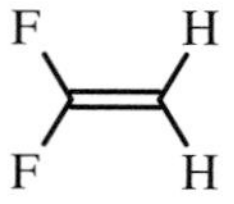 F H, F H (ethylene)	$AA'XX'$	$AA'XX'$	Topologically the same as the previous example.

Molecule	Type of spectrum in non-chiral environment	Type of spectrum in chiral environment	Remarks
R_1, R_2, R_3, C, H, H, H, H, R_4	$AA'BB'$ or $AA'XX'$	$AA'BB'$ or $AA'XX'$	Though the molecule is chiral, a fast rotation around the Ph-$CR_1R_2R_3$ axis leads to pairwise isochronicity of the aromatic protons.
R_1, H, H, H, H, R_2	$AA'BB'$ or $AA'XX'$	$ABCD$ or $ABXY$	The geminal protons are enantiotopic and there is no symmetry relationship between the vicinal couplings. In practice the spin system often mimics an A_2X_2 one.
H, H, H, H, R_1, R_1, R_2, R_2	A_2BC	$ABCD$	The methylene protons are enantiotopic.
H, H, H, H, R_1, R_2, R_1, R_2	A_2BC or A_2XY	$ABCD$ or $ABXY$	The methine protons are enantiotopic, the methylene protons diastereotopic.
H, H, H, H, R_1, R_2, R_2, R_1	$AA'BB'$ or $AA'XX'$	$AA'BB'$ or $AA'XX'$	Both the methylene and the two methine protons are homotopic but there is no symmetry relation between the relevant coupling paths.

19.5 References

[1] U. Weber, H. Thiele, G. Hägele, NMR Spectroscopy: Modern Spectral Analysis, VCH, Weinheim, 1997.

20 Short Introduction to ChemWindow

20.1 Introduction

The demo version of the commercially available structure drawing program **ChemWindow® III** [1] is used in **SpecTool®**, to draw the structure of a target molecule for estimating the carbon-13 and proton chemical shifts. Each non-hydrogen atom must be present as an individual node, i.e., an input such as CH_2CH_3 is not accepted by the Shift program.

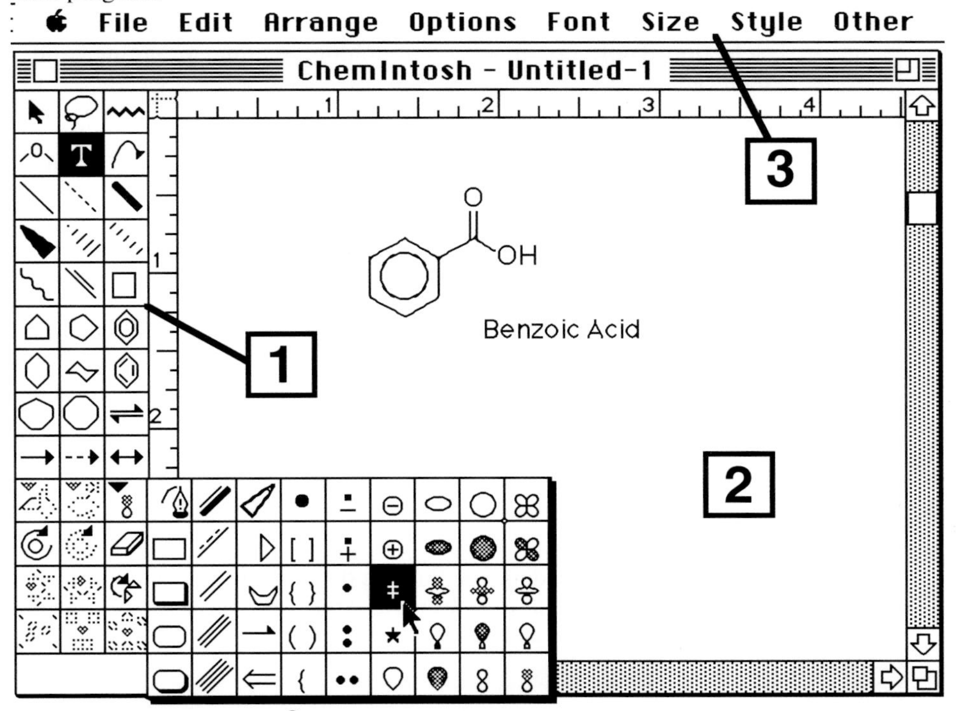

Fig. 20.1: **ChemWindow® III** user interface

In the following, a brief description of the most important features of **ChemWindow® III / ChemIntosh® III** is given. It should be noted that this demo version does not allow one to export graphics.

1. Tool Palette
The Tool Palette is a collection of drawing tools. Each tool symbol is called an icon. To pick up (or select) a tool, position the mouse cursor on the icon and click. Only one tool can be selected at a time. The bottom two rows of icons are commands. If an icon is gray, it is not applicable to the situation, such as when no object is selected.
2. Drawing Area
The same as on a piece of paper, you can draw by selecting a tool and by clicking or dragging in the drawing area.
3. Menu Bar
This stores commands by topic. To give a command, you must select it from a menu. To select a command, place the mouse cursor on the menu name, hold the mouse button down (menu appears), move the mouse to the command, and release the button. If a menu as a command is gray, it is not applicable, such as when no object is selected.

20.2 Tools

Selection Tool: For moving, stretching, scaling, dragging, and duplicating objects.
Lasso Tool: For moving part of structure without moving the whole structure.
Acyclic Chain Tool: For drawing acyclic chain structures in any direction.
Label Tool: For labeling atoms, drawing bonds, and changing bond types.
Caption Tool: For creating, editing, and designing captions and body text.
Bézier Curve Tool: For drawing mechanistic arrows and smooth curves.
Bond Tools: For quickly placing single, double, and triple bonds into the drawing area at preassigned angles and sizes.

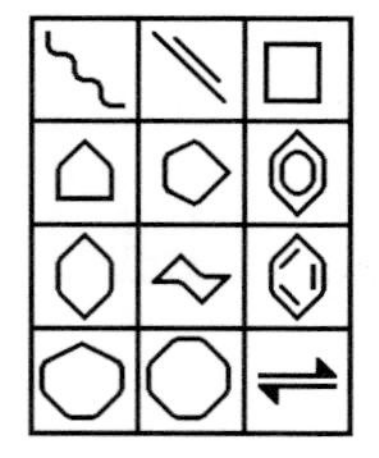

Ring Tools: For speeding you through the drawing process. They are placed uniformly in the drawing area and have preassigned constraints on angles and size.

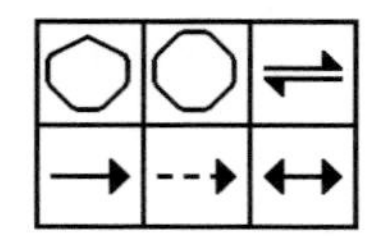

Arrow Tools: For creating reaction, retrosynthetic, resonance, and equilibrium arrows.

Template Palette: Loads in Templates.

Album Palette: Loads in Albums.

Drawing Tools Palette: For storing additional drawing tools.

Arc Tools: For drawing circles, ovals, arcs, and curves.

Eraser Tool: Deletes all or part of an object.

Flip Command Icons: For flipping an object top-to-bottom or side-to-side. 
Freehand Rotation Command Icon: Switches to Freehand Rotation mode. 
Group Command Icon: Groups objects.

Join Command Icon: Joins structures.

Align Command Icon: Aligns objects.

Pen Tool: Draws straight or curved lines.

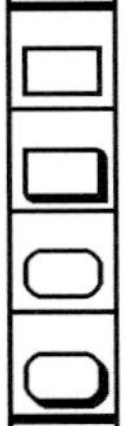

Boxes: For drawing boxes with editable corners and drop shadows.

Stereochemistry Dot: For placing dots on any point representing an atom. 
Brackets: For placing brackets that can easily be stretched and shaped.

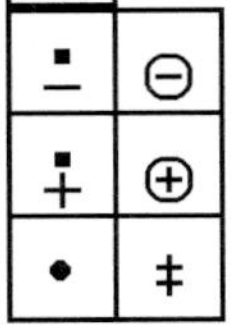

Electron and Charge Tools: For creating radicals and circled charges. Symbols can be associated with atoms.

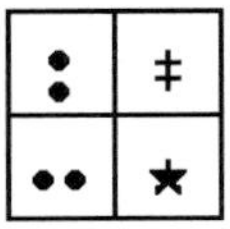

Chemical Symbols: For creating graphics that are not chemically associated with a structure. Symbols can be moved and rotated.

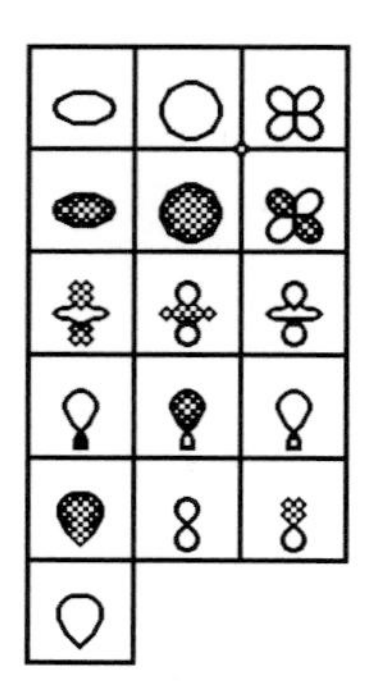

Orbitals and Ovals: For creating orbitals that can be moved, stretched, or scaled with the Selection Tool. Ovals that are used to show aromaticity in slanted rings should be drawn with the oval-shaped orbital.

Drawing Rings

Click on a ring tool in the tool palette. The icon becomes inverted (white on black) to identify the chosen tool.

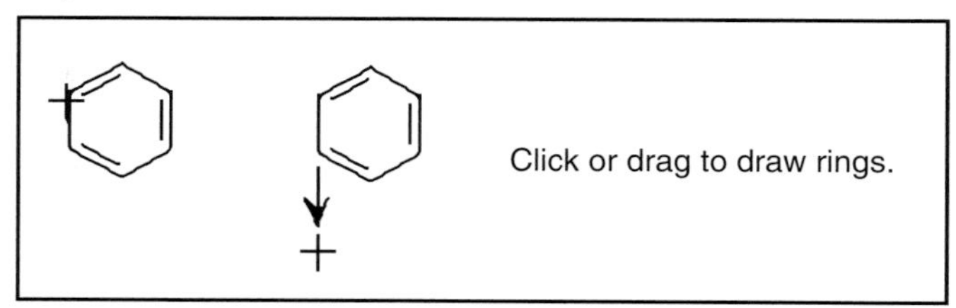

Then click or drag in the center of the drawing area. A click will draw the ring in the default position. If you drag, be sure to move the mouse an inch or two. Short drags are interpreted as aborted operations. Rotate the ring to any position as you drag. When the ring is positioned correctly, release the mouse button.

Fusing Rings

Fused rings share at least one bond. To connect parts of a structure correctly, you must first understand hit boxes. Hit boxes are small black or inverted boxes located at key points on a structure. A hit box is a visual cue that identifies the connecting points of a structure. When you draw a new object from a hit box, the software makes a smooth connection between the two objects regardless of where you actually started the drag. Chemical bonding is also understood when hit boxes are used to draw structures.

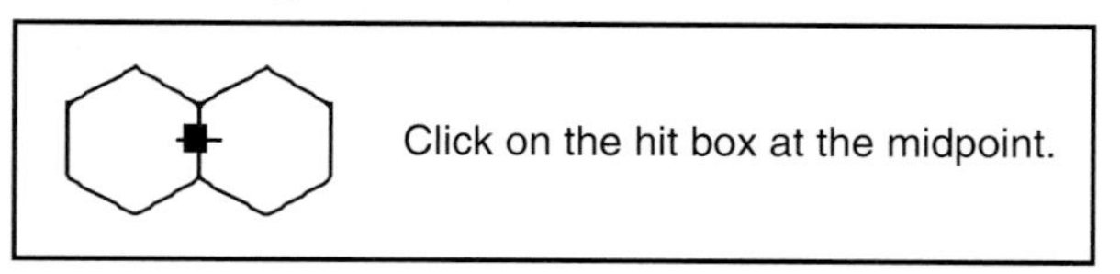

Position the cursor on the midpoint of a bond and click. The new ring will be drawn fused to the first one at the bond indicated.

Undoing Mistakes

Select the Undo Command in the Edit Menu to undo an operation. Select the command a second time if you want to undo the second-to-last operation and so on in reverse order.

Drawing Bonds

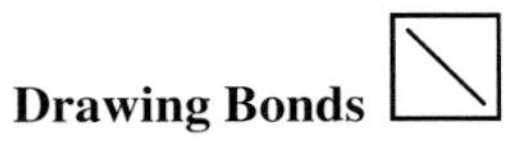

Be sure to use the hit boxes when drawing bonds. When you move the mouse cursor on to a hit box, the hit box will be highlighted.

Click on the Standard Bond Tool in the palette.

Position the mouse cursor on the hit box of an atom. Place the cursor on any atom in the ring and the hit box will appear.

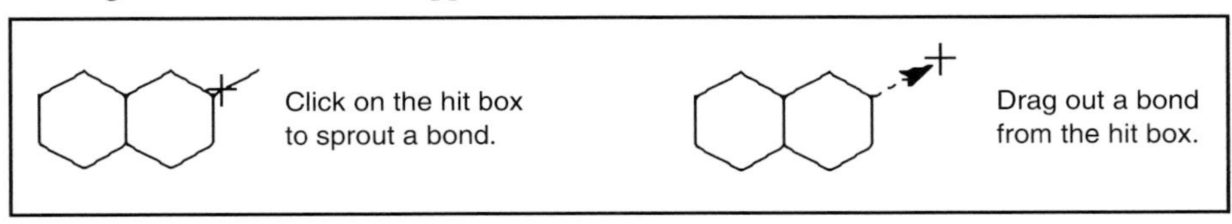

Click, or ... A click sprouts a bond out away from the structure. This is actually the shortcut way to draw a bond. For more control, try the next method.

... drag. Move the mouse in any direction and rotate the bond to any position before releasing the mouse button. The pivot atom is the connecting atom on the ring where you began your drag.

Drawing Labels

The word label is used to identify characters in a structure that represent atoms. A label may be a single letter, such as O, or a string of characters, such as CH_2CH_3. Most labels are created and edited with the Label Tool. You can also create labels using Hot Keys or the Periodic Table (not explained here).

CH_3

To label atoms, click on hit box and type.

Click on the Label Tool.

Click on a hit box. A blinking cursor will appear over the atom.

Type. The label becomes part of the object. Subscripts and superscripts are automatic.

Selecting and Deleting Objects

If you have only one object on the screen, draw some more rings which you will need to practice deleting.

Click on the Selection Tool in the palette. The Selection Tool is home base. Use it to select or identify structures to be affected by commands.

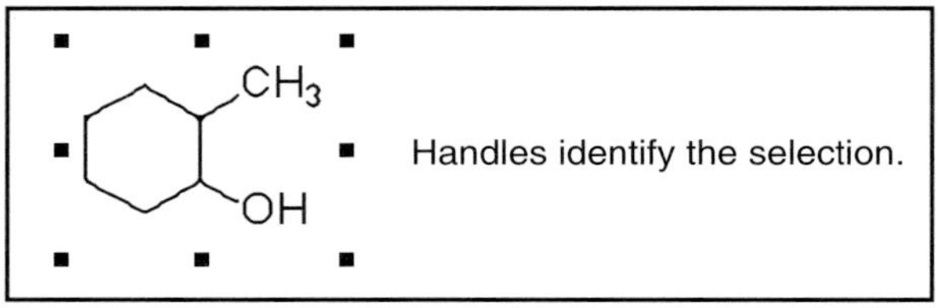

CH_3 OH Handles identify the selection.

Click on any hit box, or ... After having selected a structure, you should see eight little boxes, called "handles", arranged in a frame around it. Handles identify your selection. To deselect an object, click in any empty space of the drawing area.

... drag over a structure. This is another way to select objects. Just move the mouse cursor near a structure, press the mouse button, and drag across the structure. To select multiple items, drag over all of them.

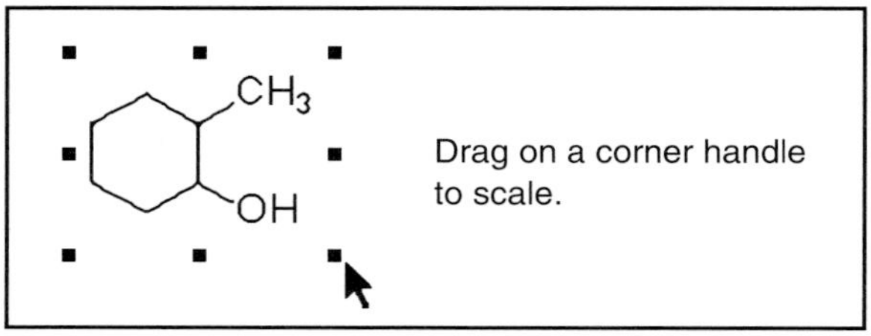

CH_3 OH Drag on a corner handle to scale.

Drag on the handles to see the effects. The corner handles scale the structure, side handles stretch it. Drag on a hit box (- not a handle -) to move the structure. When you press the DELETE key, a selected structure is removed from the document.

Fine Tuning with the Lasso

The Lasso Tool is similar to the Selection Tool except that it is used to select and manipulate parts of a structure, not the whole structure.

Click on the Lasso Tool in the palette**.**

Click on an atom or ... This is the first way to select a point (or atom) with the Lasso Tool. The selected atom is surrounded by dashed lines. You can deselect the atom by clicking in a empty portion of the drawing area.

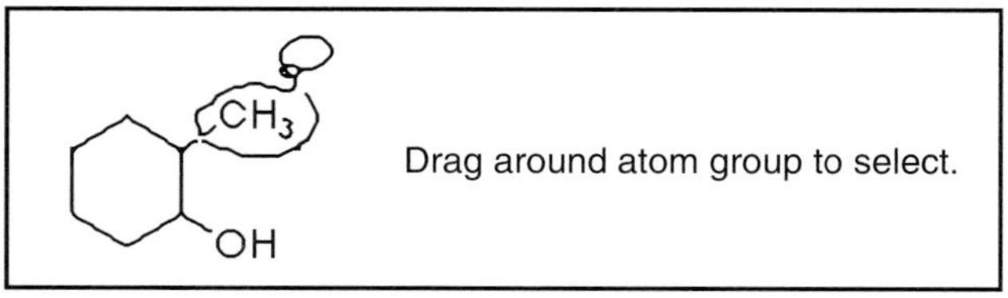

CH_3 OH Drag around atom group to select.

... drag a circle around an atom. This is the second method for selecting an atom with the Lasso Tool. You can use it to select multiple points.

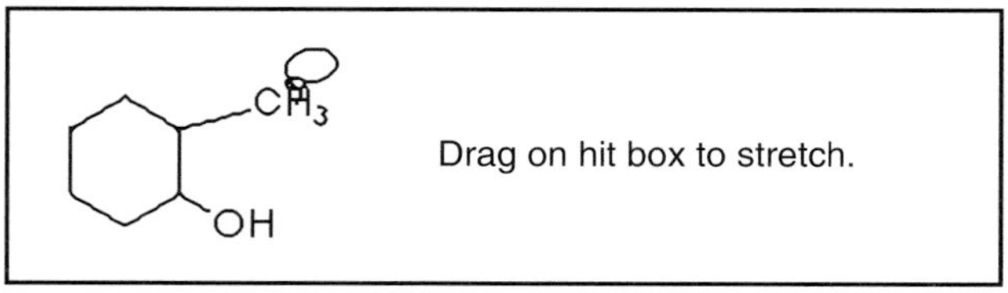

CH_3 OH Drag on hit box to stretch.

Drag the lasso-selected atom with the Lasso Tool. You must drag on the hit box of the lasso-selected atom. If more than one atom has been selected, you need only to drag on one of the hit boxes to move all of the atoms. Alternatively, you could use the ARROW keys on the keyboard to move lasso-selected points by one screen pixel at a time.

Writing a Caption T

Click on the Caption Tool in the palette**.**

Click in the drawing area where you want to start your caption.

Type. Captions may be edited using basic word processing methods. Do not use the Caption Tool to draw atom labels.

20.3 References

[1] SoftShell International Ltd., 715 Horizon Drive, Suite 390, Grand Junction, CO 81506, USA

21 Structures of Compounds

Problem	Structural Formula	Name	Page
1	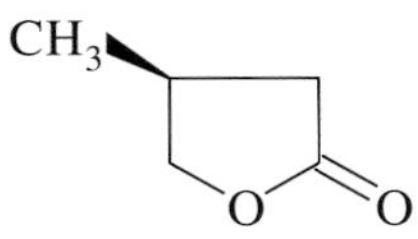 CH_3 … O … O	(S)-β-Methyl-γ-butyrolactone	1
2	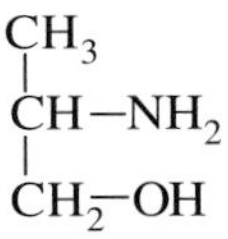 CH_3–CH–NH_2–CH_2–OH	S(+)-2-Amino-1-propanol	9
3	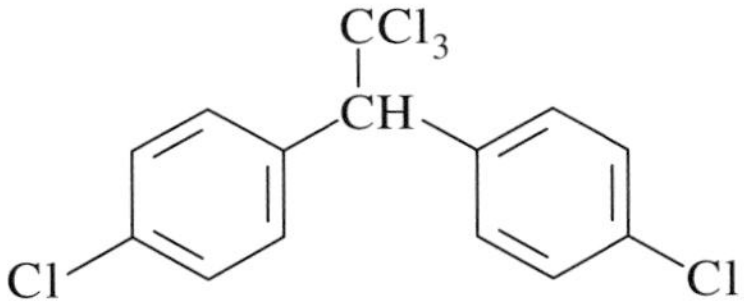 CCl_3, CH, Cl, Cl	1,1-bis(4-Chlorophenyl)-2,2,2,-trichloroethane (DDT)	17
4	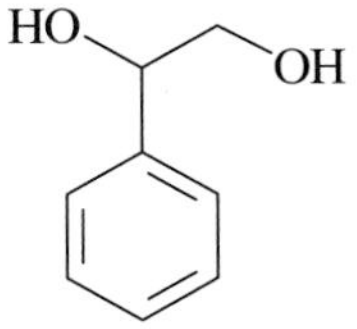 HO, OH	(±)-1-Phenyl-1,2-ethanediol	23
5	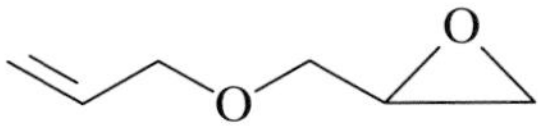 O, O	Allyl-2,3-epoxypropyl ether	33
6	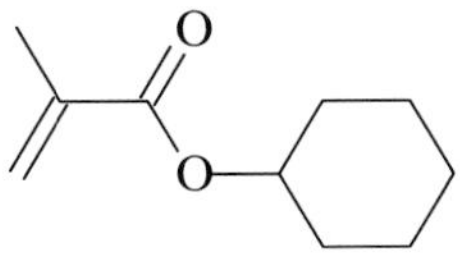 O, O	Cyclohexyl methacrylate	41

Problem	Structural Formula	Name	Page
7	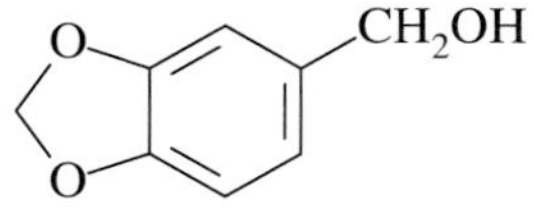 CH_2OH	Piperonyl alcohol	51
8	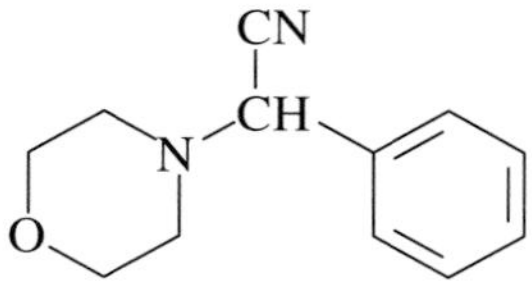 CN, CH, N, O	2-Morpholino-2-phenylacetonitrile	59
9	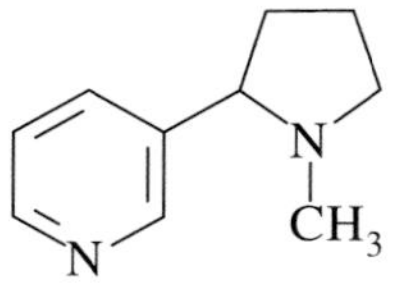 N, N, CH_3	(-)Nicotin	67
10	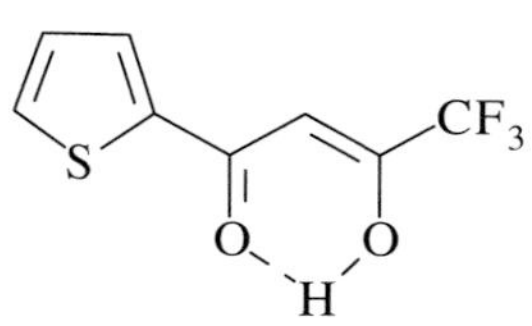 S, CF_3, O, H, O	4,4,4-Trifluoro-1-(2-thienyl)-1,3-butanedione	77
11	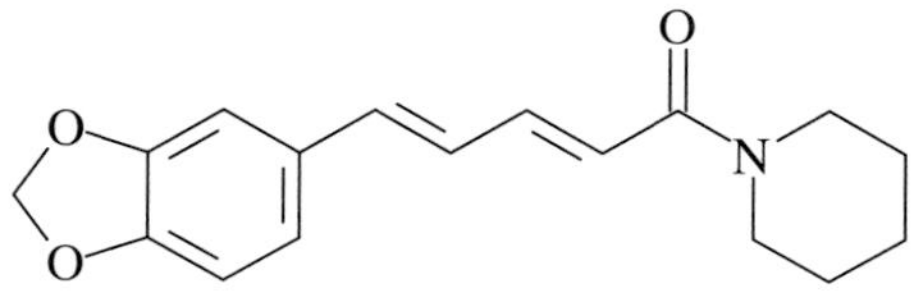 O, O, O, N	Piperine	85
12	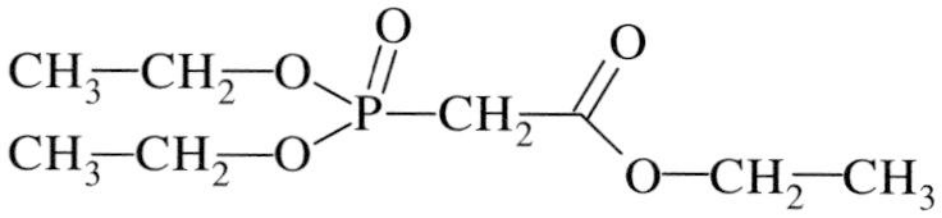 $CH_3{-}CH_2{-}O$, $CH_3{-}CH_2{-}O$, O, P, CH_2, O, $O{-}CH_2{-}CH_3$	Phosphonoacetic acid triethylester	93
13	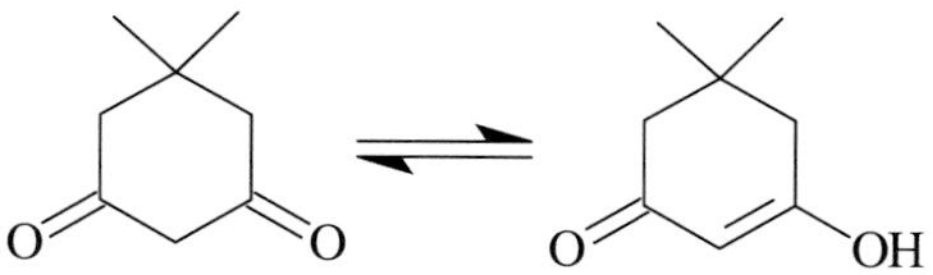 O, O ⇌ O, OH	Dimedon	101

Prob-lem	Structural Formula	Name	Page
14	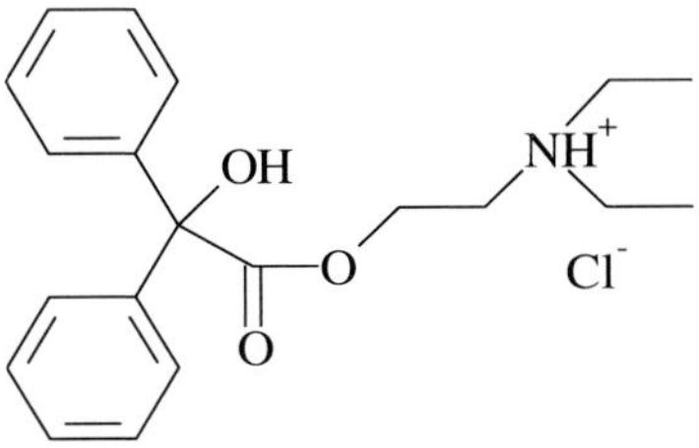	2-Diethylaminoethyl benzylate hydrochloride	113
15	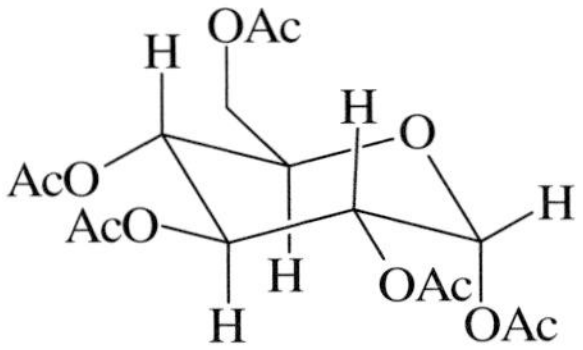	α-D(+)Glucose pentaacetate	123

Index

SpecTool 2.1

Hypermedia tools for molecular spectroscopy

- **Do you** spend time flicking through atlases of spectra?
- **Do you** still estimate chemical shifts, UV absorbance maxima or molecular fragmentations by hand?
- **Do you** wish you didn't?

If the answer to any of these questions is **'YES'**, then you need **SpecTool**.

A comprehensive electronic book on the spectroscopy of organic compounds, **SpecTool** also contains a whole range of interactive tables and programs.

SpecTool includes:

- Characteristics of the ^{1}H-NMR, ^{13}C-NMR, IR, MS and UV spectra for each class of compound
- Spectral characteristics related to structural features
- Over 800 carefully selected example spectra from all branches of organic chemistry

Optional Modules:

- **SpecLib:** 3200 spectra (>500 ^{1}H-, >500 ^{13}C-NMR, >400 IR-, >500 MS- and >1000 UV spectra)
- **Structure Generators:** tools to produce all possible isomers from a given molecular formular

Who should have **SpecTool**?

- Chemists – as an aid to structure elucidation
- Spectroscopists - as a reference work
- Lecturers - as a teaching program
- Students - as a learning aid

Technical Information for SpecTool 2.1 (October 1996)

Hardware requirements:

- **Macintosh**
  - 68020 processor
  - System 7.x
  - 4 MB Ram
  - 10 MB free for SpecTool
  - additional 25 MB for SpecLib
- **Windows PC**
  - 80386 processor
  - Windows 3.1
  - 8 MB Ram
  - VGA graphics
  - 26 MB free for SpecTool
  - additional 25 MB for SpecLib

Chemical Concepts

Chemical Concepts GmbH, Germany
Boschstrasse 12
D-69469 Weinheim
Federal Republic of Germany
Tel: +49 (0)6201 - 606433
Fax: +49 (0)6201 - 606430
Date of Information: March 1997

Yes, I would like to make my life easier!
I hereby order the items marked:

Prices (DM)

Product	Industry	Education	Students
SpecTool full version	❑ 1500.-	❑ 1000.-	❑ 195.-
SpecLib*	❑ 1500.-	❑ 1000.-	❑ 1000.-
Structure Generator	❑ 800.-	❑ 550.-	❑ 150.-
*SpecTool evaluation version***	❑ *40.-*	❑ *40.-*	❑ *40.-*
*Demo diskette****	❑ *free*	❑ *free*	❑ *free*

* 3200 spectra (including the 800 spectra in SpecTool)
** Includes diskettes and manual. The cost of the evaluation copy will – in case of purchase – be balanced against the purchase price.
***Power Point slide demonstration.
All prices exclusive VAT. Post and packing included.
Multi-user prices on request. Prices subject to change without notice.
All trademarks and tradenames are acknowledged.

Please send me the software for :

❑ **Windows** ❑ diskettes ❑ CD-ROM
❑ **Macintosh** (diskettes only)

Name
Company
Department
Address

Country
Tel. Fax
E-mail

E-mail: spectool@cc.vchgroup.de WWW: http/www.vchgroup.de/cc/

3

KETTERING COLLEGE
MEDICAL ARTS LIBRARY

QD 272 .S6 P74 1997

Pretsch, Ern¨o, 1942-

Spectra interpretation of
organic compounds